KEEP CALM AND CODE ON

CONTENTS

FOREWORD

You've probably heard the old adage that experience is something you don't get until after you need it. For those of us in the coding world, this book was written to solve that problem. But it's not what you might be thinking—this is a book about the *culture and practice* of coding rather than the coding itself.

Alex presents these lessons in a way that speaks to developers like me *then* and developers like me *now*. His book recognizes and names a set of antipatterns that are common in engineering teams—antipatterns that are not conducive to productivity and sometimes even for the health of both team and individual. It's worth asking here why antipatterns exist at all. If the subject of an antipattern is such a bad idea, why doesn't everyone intuitively understand and avoid them?

I think it's because, at first glance, antipatterns present us with a mirage—a faint, hazy illusion resembling that which we *do* want to promote. What makes something an antipattern is that it *seems* to be the shortest path between where we are, and where we want to be. But there are at least one of two qualities present in an antipattern that are missing from positive patterns. The first is that where we want to be and where we *think* we want to be are not the same destination. (With apologies to the early developers of Twitter, imagine writing a microblog when what you really wanted was a message broker.) The second (and worse possibility, in my opinion) is that the shortest path leads us directly into a gaping maw of despair that we don't see until it's too late—a "pitfall," if you will.

The tragedy of this second quality and the pitfalls it portends is that there is nearly always a way to avoid the looming disaster if only we have both awareness and knowledge of how we will arrive there.

Alex has structured this book as a series of case studies centered on the aforementioned antipatterns. There are chapters that warn about cultural antipatterns—for example, the rockstar who believes they can hack their way through all the code themselves and won't collaborate with team members. There are also chapters that walk us through the pitfalls of process antipatterns—for example, the bikeshedding team who bicker about the asinine detritus of their current feature specification rather than building literally anything of value with that energy.

Alex and I worked together as engineers at a startup in Austin a few years before the pandemic. When we were stuck on an intractable block of code for a thorny set of features, Alex would often suggest that we go for a walk. I'm not much of a walker, and I don't like to be interrupted when I'm in the zone. But he always knew when to pause and how to coax me outside. Those walks were a chance to breathe, test assumptions with one another, and remember the larger goals we shared. Alex knows how to frame and reframe a challenging issue so that his coworkers and friends come away from conversations with a new perspective.

There are several times in my life when I would've benefited from reading this book. First, as a new developer early in my first couple of roles, I struggled greatly with hubris (as many young coders are wont to do) and could have used a reality check. For example, at one of my first startup roles, I was convinced that the language (Visual Basic, for the curious) and some nebulous aspect I called "the architecture" were at fault for difficulties and a lack of productivity. (Certainly, it wasn't my lack of experience or understanding that could be at fault!)

Moreover, I was unable to ask for help when I needed it. I was not even able to recognize that I *needed* help, though I'm not sure I would have sought it out if I had.

Alex's own stories ground each case study—they are relatable looks at what it feels like to work on engineering teams today. Many developers, especially early in our careers, can become quickly enamored of the new hotness: new languages, new tools, new libraries, new techniques. In The Shiny New Objects Pitfall chapter, Alex highlights the risk of glossing over the drawbacks of a new library or language. The story Alex shares here—one in which he was evangelizing for some shiny new text editor widget and was pulled back by his product manager—offers a mildly cautionary tale about unforeseen consequences. *Keep Calm and Code On* is honest, avoids jargon, and offers readers the opportunity to reflect on their own potential pitfalls.

As a mid-level engineer, I had enough experience to know there was a better way of doing things. I had been exposed to teams with better working habits and more developed workflows—for example, I'd seen teams prioritize work-life balance and test-driven development. I'd also seen teams shortchange themselves by failing to complete a migration to a new tool or by ducking the difficult conversations required when they needed to say, "No." At that point in my career, I had enough experience to sense some of what worked and what didn't work. The problem was that I had no real mentorship—no one to help me name the problems my teams were encountering, and no one to help me work through those problems. Not only does *Keep Calm and Code On* offer readers the feeling of recognition when our work challenges are clearly named, but it extends hope that we might be able to find a path forward.

Alex is frank about the challenges confronting engineering teams and leaves us with questions to guide our thinking rather than offering platitudes and prescribing canned solutions. For example, often pressure outside our immediate development teams leads to decision-making driven by short-term objectives. In The Default to Yes Pitfall, Alex makes clear the problems when engineers end up overcommitted when implementing features or specifications that may solve *a* problem but not the *right* problem. He advises those of us who find ourselves in that situation to consider pushing back and guides us through the drawbacks when we don't. Reading this book feels at times like venting with a colleague or having beers with a mentor.

Reading early drafts of some of these pitfalls helped crystallize for me the challenges I'd so often experienced but hadn't verbalized even with decades of experience. By the time I moved into more senior roles, I had the intuition to make the right calls and (mostly) avoid the wrong ones. In fact, when leading teams

I often have the sense that we're headed somewhere I've been before. And yet, encountering familiar problems and helping others navigate through those problems are two different tasks. In naming the pitfalls, Alex offers senior developers new tools for leading teams by providing a shorthand for discussing the complex problems that arise in our work.

Ours is not a field that pauses long for reflection. With this book, Alex has distilled the most fraught moments developers will stumble through and offers us the chance to consider what he's done, what we might do differently, and how we might help others avoid these pitfalls of programming.

Keith Gaddis
November 24, 2023

INTRODUCTION

I've been writing software for over fifteen years. In that time, I've seen a lot...

I've been in the excitement of a startup being acquired. I've literally jumped for joy when our 5-person company signed its first large contract with one of the biggest banks in the US. I've been at a company that folded. I've seen coworkers laid off. I've been laid off.

I've worked at many startups, medium-sized companies, and a Fortune 100 company. I've seen when teams are organized efficiently and collaborate effectively. I've seen the infighting and politics that get ugly when teams have been pitted against each other for resources.

I've seen great code. I've seen terrible code. I've taken great pride in refactoring and simplifying complex systems. I've been the person who wrote greenfield applications that other developers cursed down the road. I've also been the one who maintained old applications and cursed other developers for their past decisions.

I've made many mistakes in many areas of my approach to software. I've overworked myself. I've promised overly-optimistic deadlines that I couldn't meet. I've convinced myself that writing more code resulted in more value, even when it most certainly did not.

But before we dive *too* deeply into all of these ways I've screwed up, know that in the end I've been able to adapt and conquer the obstacles that my career has thrown at me. In this second half of my career, I've hit my stride. I've set my ego aside and learned to listen to others. I've landed my dream job. I've been able to hold my own as I written code side by side with extremely talented engineers.

Why am I telling you all of this? Because the lessons in this book are born from scars. They're not from someone who experienced a fairytale version of a career—they're from someone who knew very little when he started and had to figure it out along the way.

But here's the good news: where I've struggled, you'll succeed. I want you to know that even those of us who aren't natural geniuses can become amazing software developers. And the better news is that unlike me, it's going to take you a lot less than ten years to hit your stride.

Why Pitfalls?

A handful of colleagues have always seemed to stand out as operating on another plane. Work in software long enough and you'll come across them, too. It's hard to articulate exactly what made these developers so effective. Part of it is that many of the struggles I found myself encountering were non-issues for them. They rarely succumbed to being surprised by edge cases, having tickets rejected multiple times during QA testing, or having their code torn to absolute shreds during code review. Those who embodied this notion had a positive influence on the entire team instead of saving all the glory for themselves.

They also all possessed a great deal of *foresight*. This foresight gave them the ability to sidestep problems entirely. It allowed them to trust their judgment and tactfully recover even when they had started going down a wrong path. Their upper hand came not from the tendency to do everything right, but to consistently avoid disastrous mistakes.

I wondered how they obtained this foresight. Surprisingly, if you were to ask many of them, they likely wouldn't be able to break down their thought process with the same grace as they approach code. From the outside, the talent appears to be innate.

For a long time, I begrudgingly accepted the idea that these individuals' talents came naturally. As I worked with more of these developers over the years, I gradually began to realize that the true source of this foresight was something beyond their raw code writing ability—the weapon they would subconsciously wield was the tactical approach they had before, during, and after code was written.

"Give me six hours to chop down a tree, and I will spend the first four sharpening the axe." is a quote by Abraham Lincoln that captures this spirit. In the same way that chopping down a tree is less about raw strength and more about the appropriate preparation, so too can your approach in software development help leverage better results with less effort.

As ironic as it is to have a book about coding without any code in it, that's precisely what you'll find here—with great intention. There are plenty of other books out there that specialize in improving code itself. Instead, this book focuses solely on your *approach*—specifically what to do and what not to do to be effective. I've chosen to concentrate on pitfalls for the what-not-to-do sections. Why? Because doing things incorrectly is where our biggest opportunities for improvement lie in this life, and the world of code is no different.

Many of the lessons here are ones that I've learned in the most drawn-out, inefficient ways imaginable. Although I am thankful in hindsight for having experienced these lessons first-hand, the vast majority of them could have been learned just as easily—and with much less pain—had I had more familiarity around them in the first place.

Make no mistake, writing the code itself is of course a requisite part of becoming a great software developer. But coding is only one piece of the puzzle. Software development rewards diligence, and having the proper mentality for software is just as critical. Knowing how to handle the tough circumstances where failure is unfolding or imminent leads to the greatest learning opportunities. Being able to navigate the pitfalls of software development will take you down the path to excellence.

What is a Pitfall?

For the purposes of this book, a pitfall is a subtle problem encountered in the process of software development. It's a common method that people use when in reality, there is clearly a better alternative. True to the literal definition of a pitfall, some of these antipatterns are more veiled than others.

It's hard to fix what we can't name. In order to capture the spirit of these recurring phenomena that teams face, you'll find names like "The Rockstar Pitfall" and "The Writer's Block Pitfall" in this book. It's less important to use "official" language when unpacking these pitfalls than it is to provide a common, memorable, and accessible language with which we can discuss them.

It's worth pointing out that there are a handful of nuances between the terms antipattern and pitfall. Some would weigh the idea behind a practice as being fundamentally good or bad before classifying it as a pitfall or antipattern (a pitfall being a bad idea from the start). Although I'm not opposed to this definition, I have been purposefully less rigid here when referring to pitfalls.

What to Expect in This Book

This book makes a handful of assumptions for the sake of simplicity.

The recommendations revolve around working on software applications and contexts with which you are most likely to work in this field. That is to say, it does not include applications in the realm of launching rockets or operating life-saving medical devices. Working on those types of mission-critical applications involves a different approach than what's suggested here, as those programs have an uncompromising level of safety as a requirement. Suggestions like moving quickly and embracing experimentation are made here with the understanding that consequences for not getting everything perfect the first time around are less dire.

Another assumption made throughout the book is generally how teams are structured and operate. For our purposes, the default makeup of the teams includes a product manager, a designer, a QA tester, and a handful of developers working within two-week sprints. As we move forward, you'll note that I use the term "stakeholder" to refer to any user who is requesting new functionality in a system. Likewise, the terms "developer" and "engineer" are used interchangeably.

Each chapter of this book covers a pitfall and includes ways to deal with it. Although the chapter order is intentional and some make references to others, they are by no means required to be read in order. I would highly encourage you to skip around to different chapters that you find intriguing.

There are questions at the end of each chapter intended to inspire some introspection around how a pitfall may affect your own career. If I may step on my soapbox for a minute, I imagine there's probably some temptation you may experience to breeze through these questions in order to finish this book sooner. But I honestly believe that pausing to really stop and take a moment for self-reflection is the *best* way to internalize the lessons in this book.[1]

Lastly, I'd love to hear about your experiences with these pitfalls—no matter how major or minor they may be. If you're open to sharing them, please email me at hi@pitfallsbook.com.

---

[1] Starting a book club with friends or coworkers takes this engagement one step further as talking through these concepts makes it that much more likely that you'll apply them later in your career.

Before We Begin: Shaking Imposter Syndrome

It's common for software developers to feel like frauds at some point in their careers—a phenomenon known as "imposter syndrome"—even if they're great at their job. Does this sound familiar? For me, this has been true many times over.

Imposter syndrome manifests itself in a variety of ways, but it generally involves struggling with confidence and (falsely) viewing yourself as a phony waiting to be found out. Your self-esteem can plummet, and you are often left feeling like you can do no right in any aspect of your job. It's an affliction that is agnostic to how much experience you have, affecting both novices and experts.

There are a variety of reasons why imposter syndrome is so common in software development. For starters, software is hard to qualify and quantify. It's not a physical deliverable whose quality is immediately obvious, nor is it easy to tell how long it takes to produce a target piece of code. Furthermore, it's difficult to accurately assess how much knowledge others on a team may have; someone with imposter syndrome may assume that everyone else has the answers that they lack even if that's not true.

Feeling like an imposter can be an isolating experience that causes a vicious mental cycle. We are certainly not going to be doing our finest work in this state of mind. Beyond writing code, feeling welcomed and comfortable on a team is an essential part of any developer's role, and imposter syndrome prevents us from reaching our full potential.[2]

Adopt a Beginner's Mindset

Beginner's mindset is a concept from the Zen Buddist monk Shunryu Suzuki that encourages someone who is developing a skill to approach it as a beginner would. Typically beginners are less harsh on themselves, as they acknowledge that they are new to an activity and thus are more forgiving and understanding of mistakes.

Recalibrating expectations in this way provides a powerful shift in perspective. The irony of imposter syndrome is that if you suffer from it, you are likely a high performer who happens to be overly-critical of yourself. Whereas imposter syndrome nudges us to constantly ruminate on our own actions, beginner's mindset acknowledges mistakes but uses them as learning opportunities.

Beginner's mindset is so effective because it promotes being open to feedback. From the first draft of code to the initial proposal of a new architectural pattern in a system, there is a hesitancy around being "wrong" and changing our minds about something. Be that as it may, software is never completely

---

[2] If this concept resonates with you, I highly recommend you to check out the resources listed at https://pitfallsbook.com/resources.

"right" or "wrong," and a beginner's mindset instills wisdom by welcoming alternatives and encouraging us to think through the trade-offs of each solution.

Imposter syndrome is so prevalent in software development that it may never be outright eliminated for some of us, but it can be greatly mitigated. Recognizing when imposter tendencies begin to take shape is incredibly helpful. Adopting a beginner's mindset can take this even further. By embracing a robust mindset over perfection, we as practitioners are much better equipped to overcome obstacles and, in so doing, build confidence as esteemed members of our teams.

The reason I bring this up now is because the pitfalls presented here are imminent dangers. As you move through the chapters in this book, you'll note many obstacles that you have faced or will face in your career.

I encourage you to resist the urge to beat yourself up over past failures or feel intimidated about what's to come. Instead, this is an invitation to adopt a beginner's mindset regardless of your level of experience. Committing to thoughtfully understanding these pitfalls and actively engaging in ways to avoid them will undoubtedly have you making massive strides toward unlocking your full potential as a software developer.

Let's get started.

PART I: MINDSET

The Rock Star Pitfall

Lone superstars are overrated and actively harmful. The best developers embrace team-work and collaboration to accomplish tasks.

I used to idolize a developer on my team who was a lone wolf. He wrote the most code and came up with the most complex systems. He would disappear into the abyss for weeks and emerge with sacred deliverables that only he could have fashioned in that period of time.

Everyone looked up to the lone wolf, and I modeled my behavior after him as I went on to work at another company. I strove to be the guy who kept track of everything about a system in my head. I took immense pride in cranking out large features in isolation and adored the idea that others would have to come to me for deeply nested voodoo knowledge that I alone possessed.

Then, I left for vacation for a week. As you can imagine, it was a painful experience for both me and my team—all because I had my hands in many areas of code and didn't make any effort to develop sustainable code and artifacts.

I never relied on having any documentation. My team had been conditioned to just reach out over instant message with questions, and I would reply right away. When making larger changes in the codebase, I outlined patterns so intricate that they inevitably led to misuse. Neglecting this support meant that others who would attempt to help while I was unavailable were set up for confusion.

The code I worked on rarely had dire bugs, and because of this, I found it quicker to skip writing tests altogether. Nobody complained since I would fix issues myself anyway in the event that something came up. Unfortunately, the code you write in the office doesn't take a break while you're on vacation, and I ended up being pinged during my trip when a cryptic error showed up that was related to something I had worked on.

My work-life balance was abysmal. After this first vacation debacle, I found myself checking in on vacations—the one time I was supposed to be disconnecting entirely. But vacations weren't the only time my quality of life suffered; it did so in my day-to-day as well. I remember going out to eat with a friend for breakfast one morning before work and checking my phone non-stop, even rudely pausing mid-conversation a few times just to reply to a few questions from coworkers.

I didn't learn my lesson by the time I left that company. Absurdly, I found myself scrambling even after I put in my notice that I was leaving for a position with another organization. I still cared deeply about the product and my coworkers, so I frantically tried to make up for lost time by brain dumping as much documentation as I could into a huge manifest. There are many problems with this approach, but the biggest one was that nobody was ever going to read it because it was so overwhelming and haphazardly written.

The bottom line is that in an effort to help my team, I actually hindered them. I certainly helped make things happen *fast*, which is what teams may need from time to time. But my lack of diligence ended up making everything I touched brittle, and it became clear that I sacrificed resilience and cross-training so that I could deliver more features in the short term.

I had turned into my idol, the lone wolf developer, only to realize that I had chosen the wrong role model.

The Perils of Rock Stars

Software development can be a funny profession. In no other

field can two similar roles be titled by something as dry as "Software Engineer IV" to something as outlandish as "Ninja Code Wrangler."

Hiring gimmicks aside, hero worship is something to which software teams are not immune. Perhaps it stems from the Hollywood trope of a programmer who functions as a one-man army, feverishly coding throughout the night in a caffeine-induced state of transcendence, all to emerge with pristine code that could have only been forged under such grueling conditions.

One of the terms used to describe such a person is "rock star." But although these so-called rock stars undoubtedly have their own talents, the practices portrayed by this stereotype are at best short-sighted and at worst they set the team up for imminent failure.

Breaking down the behaviors of a rock star reveals how they affect a team in the long term. Suppose a developer abides by the Hollywood example of staying up all night to work on a feature on a tight deadline. Let's look at the qualities they're embracing.

Writing code hastily. Typically code that is written quickly is written carelessly. This carelessness may manifest in the form of bugs, poorly defined interfaces and abstractions, or verbose code with hard-to-read variable names. With a discipline as intricate as software development, elegant code isn't born under rushed conditions.

Sloppy code is distinctly harmful when it's relied upon by many areas of an application. It's common for startups to embrace Facebook's old motto, "Move fast and break things," but when it comes to architecture decisions that are core to a product, spending more time upfront making thoughtful decisions far outweighs the pains of supporting a malformed solution for years to come.

Working in isolation. "Knowledge silos" occur when only one person on a team has an intimate understanding of how a system or process operates. It's enticing for a team to have a go-to person for a select area of an application because they can fix or extend that code the fastest. This is especially true on a team that is short-staffed, and redundancy in code comprehension is viewed as a luxury in place of a necessity.

Having knowledge silos inevitably comes back to bite a team. In the world of code, it's the equivalent of putting all your eggs into one basket. The more a team relies on one person to know a part of a system, the more painful it is when that person eventually leaves. Sometimes a team will luck out and realize this when the expert is on vacation, but more commonly it's not apparent how crucial this person was to a system until they have already left the company.

Delivering code all at once. Code that's shipped in one big piece is harder to review and more inclined to contain bugs.

It also gives the product team fewer opportunities to preview behavior. See The Code Delivery En Masse Pitfall for further discussion on this practice.

Collaborators Over Heroes

A developer's success shouldn't be measured by how many miracle projects they can pull off via their own grit. Code that only they understand is no badge of honor. In fact, if your team's greatest asset is a rock star developer who is viewed as a miracle worker, you're relying on miracles. That's a tragic way to produce software. Your team can and should thrive without rock stars. The best leaders in software are collaborators who help amplify teams rather than stealing the glory for themselves.

Collaborators strive to share knowledge. They know that no individual will be around forever to explain things. They recognize that articulating the intent and structure of how an elaborate system works is immensely more beneficial to a team than cranking out feature after feature that only the author understands. They know when to clarify behavior in code comments or when to make sure a feature has corresponding documentation. They embrace pair programming and make the effort to explicitly have other developers pick up tickets even when they themself could have done the ticket faster and easier. They play the long game even though it may be more demanding.

When it comes to impact, the rock star will always fall short of what the collaborator can achieve. The rock star's biggest contribution is writing lots of code, but there is a limit to how much code a single developer can write. There is no limit, however, to the impact a developer can have when they focus on trying to make the entire team better.

Closing the Loop

You may think I regret ever becoming the lone wolf whom I had once looked up to. But in some ways, I think it was inevitable; seeing the rock star approach portrayed in movies made it seem commonplace. Leading with the wrong approach solidified the lesson I needed to learn at the time—being able to deliver features quickly and autonomously is a useful skill to have, but proceed with caution. It's easy to make that your identity in an attempt to become the hero. The true heroes are those who realize the limit of one person's contribution, lay a foundation for any developer on the team to be successful, and amplify those around them in pursuit of a shared goal.

Code comes and goes. It gets refactored and deleted. Lessons taught to other developers live on. They survive longer than any feature or project. They are passed on and help members of other teams.

Forging your path:

- What is it about the Hollywood trope of a rock star developer that's so appealing?

- Have you come across a rock star developer before? What impression did they leave on the team after they left?

- What's a skill you could develop in order to become a better collaborator?

The More Code Is More Value Pitfall

The finest code in the world is futile if it doesn't solve an actual problem that users have.

My product manager and I worked in lockstep. We had reached that rare state of collaboration where it felt effortless to solve one problem after another. As the engineering team lead at the time, I was elated with how the project was going.

Our task had been to build a brand new user survey feature for our site. We were excited to build it from the ground up and felt it had great potential to help our sales team convert more leads than ever.

Engineering-wise, features were actually being completed ahead of schedule. I'm proud to say that the code our team delivered was well-written and well-tested. We even managed to work on some nice-to-have stretch goals, too. Users were able to add lots of unique customizations, and our design team was pleased that we'd managed to include some fancy animations to polish the experience. We set out to build the best damn survey feature we could, and that's what we did.

Finally, we released the feature into the wild, eager for it to boost sales. Then we waited. And waited. And waited. It turns out that we built a fantastic feature... that nobody wanted or needed.

The Perils of Equating More Code with More Value

As a software developer, it's easy to consider writing more code as delivering more value. After all, if your job is to write code, surely more is better? Unfortunately, that's not the case. In fact, less code is almost always better. A feature that is implemented in 10 lines of code is likely more readable and less error-prone than the same functionality written in 100 lines.

The More Code is More Value Pitfall occurs when a developer thinks that delivering more features is the only way to add utility to an application. Its underlying implication goes beyond strictly counting the number of lines of code you're writing and drills to the core of why software is written in the first place.

Developers engaged in busywork aren't adding substance to their product. Nor are developers who craft immaculate algorithms for an area of an application that is rarely used and solves a low-stakes problem.

Real value is created when an application solves the needs of its users, not when it has more features. Judging an application by the number of features is like judging a restaurant by the number of items it has on its menu. Having 200 options may seem impressive to some at first glance, but the truth of the matter is that the restaurant will never be able to deliver excellent quality on all of them and will eventually spread itself too thin. This means that the restaurant's operations suffer in the process—from supply restocking to wait times, and everything in between. Options *sound* good, but we really go to restaurants to choose from a handful of select menu items that are skillfully prepared. In the same way, users of an application only need a limited set of functionality but need it to work well.

This can be a tough mindset to adopt in organizations where performance is measured by the number of features shipped and not the real problems being solved. Moreover, it's challenging to come to a consensus on what constitutes value without some sort of framework in place. If the goal is to increase page views and there is an analytics framework that is already being used, figuring out if that goal has been met is incredibly straightforward. If the goal is to make a page "easier to use" then there is more discussion that needs to occur in order to figure out what categorically constitutes success.

Truly appreciating this pitfall is difficult; experience is what solidifies its merit. For some people, this means being burned by writing lots of code that is ultimately unused. For others, it takes being on a team that embraces constructive product research habits (like measuring site usage and regularly performing customer interviews) to come to the realization that writing lots of code is not the same thing as delivering a usable feature.

Consider this scenario: there is an application where users can download a report to view all of their activity on a website. For our purposes, assume this feature was thrown together hastily, resulting in a slow report that takes a full minute to download.

The team responsible for this report notes that it's not being used too frequently and makes reasonable assumptions as to why: perhaps the download is slow and/or there isn't enough useful information on the report itself. Tasked with the objective of "improve reporting," an eager developer dedicates a full two-week sprint to updating it. The developer approaches the problem with diligence—writing tests to ensure the code is robust, adding new fields to provide more information to the user, and even optimizing the queries to make the report download in a few seconds instead of a minute.

It sounds like the developer has done everything right to make the report better than ever. The advantages of the new version are undeniable—the code is well-tested, there's more information available, and the report is faster.

The real issue has less to do with *how* the developer approached the problem and more to do with *why* they approached it. The team eagerly jumped to the conclusion that the report is undoubtedly needed in the first place when, in reality, they lack a clear problem to solve. "Improve reporting" isn't a specific request in the same way that "report the customer churn rate to the sales team each month" is.

Maybe the report is only accessed by a handful of users who only need it a couple of times a year. Maybe it's pertinent to more users, and only the performance needs to be improved. Maybe even the people who download the report don't wholeheartedly trust its figures, and the report itself could be removed entirely.

Gathering more insight into why and how the report is used in the first place would help the team determine the appropriate approach to the problem. Having one developer spend one sprint on one report probably isn't the end of the world, but consistently approaching problems in a naive manner will unquestionably lead a team astray in due time.

The bottom line? Context matters, and there is an inherent opportunity cost with software development. Every new feature comes with the concession of forgoing other features that could have been built in that time. Just because a feature may provide some utility to some users doesn't mean it was the right feature to build.

Helping Find Value

The reporting example is contrived in that it conveniently assumes that the rationale is lacking for updating the report in the first place. That example isn't to say that software developers should perpetually assume the responsibility of seeking out justification for every aspect of every feature being built. Rather, their role is to support the team overall by employing the following types of techniques:

Help the product team gather insight into how an application is being used. This can come in the form of adding analytics to an application so that the team better understands its practical use. In other cases, it could mean advocating for and watching user interviews in order to acquire more insight into a problem. Either way, a developer doesn't necessarily need to be the person executing these strategies, but they should at least be familiar enough with different options and their trade-offs such that they can help lend insight to a product team when researching solutions.

Question when a problem may not be solving a real user pain point. It's a perfectly reasonable ask to want to know more about why a team is solving a given problem. In fact, blindly spending time and effort on a feature that's never going to be used is a *disservice* to a team. On a healthy team, this degree of discourse is actively encouraged. On an unhealthy team, questioning why something is being done should be approached with tact. In the world of software, just like in other industries, receptivity varies and often depends greatly on power dynamics.

State the problem that a feature is solving for users. This keeps the focus on the pain point as opposed to a specific solution. A value statement could be as basic as a one-sentence bit of background on a ticket. You can solve problems in multiple ways, but only discussing features in terms of solutions is a form of tunnel vision when building a product.

It's natural to fall back into autopilot mode and assume that everything that makes its way into a feature request has some innate justification. The important thing is to ask the question when its answer isn't immediately obvious, as the software development process isn't cheap. There is a very real cost to this lifecycle: designing a feature, writing code, testing it, deploying it, and verifying it afterward all take time and effort.

Make verification a part of the process. It's all too common for teams to spend significant time upfront on a feature in terms of how it is constructed, only to release it into the ether and never think about it again (see The Done Upon Delivery Pitfall).

Skipping the verification process may afford temporary comfort, but the underlying issue still needs to be addressed at some point.

This strategy is well suited for features that address problems with experimental solutions. Suppose a team is attempting to reduce the number of support tickets for an application by a target percentage, and the customer support team frequently fields requests regarding confusion around different pricing plans. The engineering team may be tempted to push out a redesigned pricing page and then continue on to other tasks. If they never follow up and measure the support tickets related to pricing confusion, they don't actually know if they've solved the problem they were setting out to solve in the first place.

Test and follow up with your feature if you want to provide maximum value. Even attempting to measure the effectiveness of changes within a system is a significant first step. Teams should gather as much data as they need to answer the question "Did we solve the problem we set out to solve?" after features go out. Only then can issues be resolved with confidence.

The complementary implication of this approach is arguably more insightful—-problems in which a team is less confident should be allotted less time until they can be further *justified*. Otherwise, you end up where we did with our user survey feature; we wasted lots of time building a feature that we weren't confident in.

Closing the Loop

Looking back on the user survey feature, one possible outcome would have been for me to point the finger at our product team and blame them for their lack of validation. But that's not how software development works (or at least not how it *should* work).

I knew as well as anyone that surveys were a large feature that we were building in isolation. It was wholly within my power to advocate for building a minimum viable product and gathering user feedback first. Likewise, part of my role in the development of the feature should have been discussing how we could measure and validate usage.

It's not bad to start going in the wrong direction when developing a solution. But it *is* bad to barrel toward an outcome without ever pausing and checking to make sure you haven't lost your way.

Forging your path:

- When was the last time you wrote a big feature that ended up being discarded? What could you have done to help validate this ahead of time?

- Is your team open to the idea that less code could result in a more useful application? If not, are there any examples of features you could provide that took a lot of effort but resulted in low usage?

- What new practices could your team adopt to make value a guiding principle?

For more information on how to create value within the software you build, check out https://pitfallsbook.com/resources.

The Hardcore
Mentality Pitfall

Self-care is more important than an unwavering dedication to work.

It was 10 PM, and I was nearing the end of an excruciatingly long day. The task ahead of me was simple, but I could hardly muster the energy to take those few dozen keystrokes I'd need to finish it.

I had burned myself out into a state of stupor; that work in front of me could've been done in a painless half hour if I had just stopped for a break and gotten a good night's rest. Instead, I trudged through it in an attempt to be resilient.

I didn't recognize my own lack of productivity because I was too overworked to think clearly. I had overworked myself for weeks on end by that point. My mind and body were giving me clear signals that it was time for rest, but instead, I went further down the wrong path of grinding out feature after feature on the roadmap.

My singular focus was building sweat equity in the currency of the number of hours I could work. After all, I was working at a startup, and this was what I'd signed up for. Startups move quickly and deliver products at an uncomfortably fast rate. If we threw more hours toward a problem than our competitors, there's no way we could lose to them, right?

Wrong.

Not all hours are created equally, and hurling hours at problems without any additional tact negates the idea that there are ways to work smarter rather than harder. I was too busy being hardcore to stop and think about how much more productive I could actually be if I took more breaks to recharge and brought my A-game to work each day instead.

This is especially true in domains that require a high level of cognition. I can't imagine choosing my version of hardcore-ness when seeking out professionals to do a job. For instance, I wouldn't prefer the lawyer who's been working tirelessly for fourteen hours straight to make an opening statement for me in a case, and I *certainly* wouldn't want a sleep-deprived and overworked surgeon operating on me. Why on earth would anyone want a developer who's worked themselves to the bone to be responsible for delivering quality software?

The Perils of a Hardcore Mentality

When it comes to being a software developer, being "extreme" or "hardcore" can be a double-edged sword. A hardcore developer embodies positive traits like patience, persistence, and focus. Someone who works nonstop for ten hours on a single problem without interruption undeniably has grit and determination. The question, though, is whether or not this is a practical way to approach *every* problem. (Spoiler alert: it's not.)

There is a time and a place for being hardcore. Say a website is completely down or there is an accounting bug in the system that needs to be fixed on a tight deadline. These types of issues require immediate attention, and the price of some temporary discomfort merits addressing a dire need within a system.

However, the reality is that the vast majority of problems aren't truly urgent, and undertaking *every* problem in a hardcore manner fails to optimize the approach. It's better to address a problem from a position of being rested and energized than to be drained from pushing yourself to the limit. That said, in a cruel twist of irony, the times we need a break the most are often the times we are most reluctant to take one.

I'm not here to lecture you on diet and exercise. Instead, consider this an invitation to internalize that the mind and body are fundamental pieces of hardware to maintain in software development. Even setting aside the quality of life aspect and considering self-care from a purely utilitarian standpoint, disregarding maintenance results in worse code.

Balance as a Competitive Edge

For many, jumping to a hardcore mindset is simply following the gut reaction approach to a problem. Deciding to work without distraction until an issue is resolved is an easy formula to follow. At first glance, it does seem reasonable that anything that isn't directly helping to solve a problem can be viewed as an unnecessary distraction.

The problem is that humans are messy. We're not mere cogs in a machine whereby our output can be flawlessly predicted when handed a task. Instead, work output wholly depends on context.

This context could be anything from the number of hours slept to being well-fed to even general mood. Since so many factors affect your output, it doesn't make sense to attempt to optimize productivity by turning the dial all the way up for one factor only to be lacking in others. Taking the approach of, "I'm going to finish this task before the end of the day no matter what," is foolhardy when you're sleep-deprived, hungry, or overwhelmed to the point where you can't even focus.

Finding a healthy balance between what's needed on a given day leads to consistent and successful long-term results. Even acknowledging the current context can help plan for a task more strategically. For example, a developer who takes stock of how they are feeling can plan accordingly. If they are sleep-deprived and hungry, they can recognize that in truth the smoothest way to finish a task is to have a quick rest and bite to eat. In a similar manner, a developer who is overwhelmed and hits a roadblock on a perplexing problem can find that going for a walk and clearing their mind has a better payoff than struggling at their computer in turmoil for the sake of formality. These examples of course sound obvious in retrospect, but it's amazing how many developers get sucked into their work and neglect simple practices that help them approach work with a clear mind.

To some, being balanced may come across as boring. Admittedly, a routine of consistently getting a good night's rest, avoiding overwork and burnout, and taking breaks throughout the day doesn't fit the image conjured for the Hollywood trope of a developer (as we saw in The Rock Star Pitfall). That strange, misleading badge of honor that some people strive toward as a result of working a sixty-hour week won't ever be attained by someone who embraces balance.

What's not boring are results, and prioritizing self-care when writing software delivers results. If it helps, try considering self-care as a way to constantly tip the scales in your favor. A balanced developer embraces their mental and physical wellness, giving them the increased capacity to be tactical about approaching problems.

Closing the Loop

It wasn't until I moved away from that startup that I reset my expectations around what a workday should be like and stopped going all-out. In addition to feeling better physically, I found myself delivering consistently better code when I was taking care of myself. The code I wrote was more well thought out, better tested, and better documented.

But my biggest realization wasn't just that self-care is important. It was that I was focusing on the wrong things. If you're a developer who is trying to churn out the most features possible, you're probably not thinking about how a team can be more strategic.

Strategic leaders are infinitely more valuable than those who are just good at execution. Who sounds like a better leader to you? The boss who's really good at answering dozens and dozens of customer service emails by hand every day? Or the boss who creates a template for the most common questions to allow other members of their team to answer emails? The first boss is hardcore, yes, but that doesn't mean their approach benefits the team.

In short, working smarter beats working harder.

Forging your path:

- Coding or otherwise—what areas of your life could use more balance?

- Think about the last feature you wrote that you were truly proud of. What were the conditions like when you created that code?

- Is there anyone you know who has an exemplary work/life balance? What do they do to achieve this perception?

For more information on creating a healthy work/life balance, check out https://pitfallsbook.com/resources.

The High Pain Tolerance Pitfall

Tolerating painful practices takes a compounding toll on teams.

None of our team's practices sounded too cumbersome... until I found myself explaining them to a new teammate.

I would tell him things like, "Oh yeah, we see that error all of the time. Just follow these five obscure steps to fix it." Or "That error notification in our chat channel doesn't mean anything; just ignore it."

We even had a handful of very automatable tasks that our rotating on-call role would run each Monday. What could have been scripted away with a few hours of effort was instead draining precious minutes from us on a regular basis.

I reaffirmed to our teammate that there was no cause for concern. He never outrightly proclaimed that these tasks were inane, but he did raise an eyebrow every now and then. Every time I explained a wonky, backward practice, I found myself ending the explanation with, "Don't worry, you'll get used to it."

Then, a few months later, another person joined our team. None of the existing problems had been fixed in the interim—in fact, workflows had only gotten more complicated. As you can imagine, their onboarding process was just as painful.

I overheard the two newest engineers talking when one repeated my phrase, "Don't worry, you'll get used to it." It was at this moment that I realized we'd screwed up; we had trained our team to tolerate pain instead of solving problems.

The Perils of a High Pain Tolerance

Those with grit are no strangers to the idea of having a high pain tolerance. The profession of software development eventually tests the mettle of all developers. Projects grow expansively with more and more features rather than reaching a point of stability. If not watched carefully, technical debt subtly emerges in this growth process—from a convoluted environment setup to unoptimized code to unmaintainable architectures.

The High Pain Tolerance Pitfall occurs when subpar status quo practices live on longer than they should due to the perceived difficulty of just *fixing them*. Consider, for example, a scenario in which a team has a report that needs to be manually generated by a developer once a week. No one is bothered by it too much because it "only" takes fifteen minutes of the developer's time. It's easy for the team to sweep this problem under the rug and indefinitely incur the cost of that fixed amount of time per week.

The mistake—and where this pitfall arises—is when similarly inefficient practices unknowingly creep into a team's workflow. Presumably the team is okay with a developer spending those fifteen minutes on the report, but what about if the business stakeholders need "just one more report"? That kind of mentality can be a slippery slope, and before the team knows it, they're spending hours each week on a task that has largely slipped in under the radar.

Beyond a busier schedule, what happens when the developer who runs the weekly report goes on vacation or leaves the company? When a team tolerates hastily written code, how is that going to affect a new team member's onboarding and fixing bugs in the future?

A team with a high pain tolerance gradually becomes less and less productive. Instead of being able to focus on the task at hand, they're marred by a series of annoyingly small hoops to jump through. Think about it this way: the road filled with potholes is still usable and will allow someone to make it to their destination, but the path will never be as fast nor effortless as the road that is smooth and properly maintained.

Easing the Pain

The most fundamental step toward navigating this pitfall is having *some* sort of system in place to address it. Let's look at how you can help your team identify, justify, and prioritize pain points.

A Group Effort

The act of finding and weeding out painful processes needs to be instilled at the team level in order to function. Since technical debt is a natural byproduct of software development, having one or two champions who actively work to reduce debt isn't enough. Even a remarkably productive engineer won't be able to keep pace with a large team.

Acknowledging shortcomings within methods of operation and proactively looking to fix them is the sign of a healthy team culture. Consider the opposite scenario: a team that doesn't look to improve surrenders to mediocrity.

In the spirit of improvement, one opportune time to capture technical debt as a ticket is while working on a feature that interacts with a challenging piece of code. Adopting the "leave it better than you found it" mindset is a healthy take on how to constantly improve a codebase. Another opportunity for reflection is during the team's retrospective. This works well for identifying previously untracked requests (like the once-weekly report from our earlier example).

For system-wide architecture, arguably the best time to gauge the approachability and coherence of a system is when a new team member joins. Their fresh eyes provide a unique perspective regarding which pieces of a system best stand to reason and, more importantly, which are the most confusing.

Looking to the Future

It's easy to say, "This process is painful, but it's not worth fixing right now." But quickly jumping to such a conclusion obscures what truly makes it painful in the first place. A high pain tolerance allows for inferior processes to live on in the short term, but surrounding factors don't have to change much to make a big difference in how detrimental a process *could* be.

When a confusing code architecture exists in a system, its consequences will vary from team member to team member. New developers joining the team are more susceptible to making mistakes while working with it. If upcoming features are going to modify or interact with this code, its relevance increases and it should be reassessed accordingly.

Each fix for a bug that occurs due to the confusing nature of a particular piece of code is unfortunate. That time would have been better spent writing clean code in the first place. Of course this can't be predicted easily, but seeing related bugs come in should be an indicator that other bugs have yet to be discovered as well. Each bug is a piece of evidence when building the case that the team needs to take more time to refactor.

The beauty of considering the holistic impact of a process is that even those with a naturally high pain tolerance can contextualize the effect of a costly subpar process. Seeing a dozen bugs crop up because a feature was rushed helps build the case for spending more time on the next feature.

Forming a Plan of Action

If technical debt constantly loses in priority to new feature requests during sprint planning, there's no incentive for a developer to capture the tasks in the first place. It's frustrating for a developer to know that there is potential for improvement but not have a means to make change within their team's mode of operation. Likewise for stakeholders, it becomes frustrating to hear that a supposedly trivial request is going to take an order of magnitude longer than they expect because the related bit of code is engulfed in technical debt. In the long term, nobody wins by ignoring pain.

In order for your team to be successful in addressing painful practices, you need to commit to taking action on a regular cadence. The specifics of the cadence are up to the team—it could be 10 percent of a sprint's capacity, a rough handful of tickets each week, or even having a specialty sprint that solely consists of technical debt in lieu of feature work. The right balance that works for your team can be discovered in due time; the meaningful part is taking action regularly.

Closing the Loop

It's common for people to grow used to painful practices when they're on a team for a long time. After a while, it becomes harder to squash every weird bug or behavior that comes your way. It's easier to apply the band-aid fix and carry on. But software systems need regular maintenance in order to prevent small issues from causing detrimental effects in the long run.

I've grown to realize what a blessing the onboarding process is because it provides a fresh perspective from a new developer. It's the perfect time to re-evaluate practices to which our teams have grown accustomed. By digging a little further into *why* we entertained these annoyances, we're able to transform pain points into systems that are more resilient and approachable than ever.

Forging your path:

- When a system's current implementation of a solution is messy, ask yourself:

  - In an ideal world where we started solving this problem from scratch, what would a solution look like?

  - What would it look like for this problem to be effortless?

- Time is just one factor when considering if a task should be automated. Next time you're debating how to handle a painful manual task, consider other factors besides time savings. For instance, will automating this task help improve reliability?

- Are there any examples of areas of your system that are much harder to work with due to technical debt? Is there a plan to address these areas?

The Unbreakable Rule Pitfall

Most coding rules can and should be broken
from time to time, but breaking them requires
careful thought and justification.

There was a long time when I thought that becoming a senior
software developer meant simply learning the most rules and
following them. Surely if I reached a certain threshold of rules
that I had memorized, I would be able to avoid any mistakes
with software, right?

Well, not so fast.

I was once in charge of adding a new page to our site and
made it my goal to incorporate every time-honored rule possible. Every change had to abide by the single-responsibility principle, the rule of "don't repeat yourself," and so on.

The resulting page was... ok. But I had overlooked the fact that the page was now completely different than every other page on our site. Dozens of other pages that already existed in the same application all followed an existing structure from which I was deviating. Sure, my page worked fine and acknowledged good coding practices, but any other developer who looked at my page would need time to ramp up and learn what I was attempting to accomplish with my new patterns. My quest for perfection left the system less approachable and more confusing overall. Ironically, it's hard to codify code.

The Perils of Unbreakable Rules

Rules can be powerful aids. Take the timeless code-writing motto "don't repeat yourself" (aka DRY). By encouraging that the same code is not written twice, this rule provides clear direction for novice and experienced developers alike.

The problem with these types of rules occurs when they are taken as a decree that cannot be broken under *any* circumstance. Of course, in most cases, it is indeed right to avoid code duplication. But many factors like readability, searchability, and consistency all play a role in determining what the author thinks will result in the quality code. Dogmatically following the DRY principles in every scenario is a sure way to obfuscate code.

On the flip side, nailing down these exceptions can be tricky. One developer's, "I'm choosing to avoid DRY here and am okay with duplication because the resulting code is more readable," may be another developer's, "I'm using DRY here so that the code is more concise." Code is subjective, and anyone may unknowingly abuse this quality when justifying whether a rule should be used in the first place.

To either abide by or knowingly break away from a rule is a choice that should be made purposefully. The best developers know how to walk this line carefully. They understand that every situation has context. They lean on teammates to solicit outside opinions to curb personal bias.

Guidelines Over Rules

In a world with an overwhelming number of rules to follow for software development, it's extremely advantageous to build a mental toolkit of different approaches. Why? Put simply, the developer with a limited collection of strategies at their disposal is akin to the carpenter who only owns a hammer. To them, every problem is a nail. On the other end of the spectrum, the carpenter who owns many tools but lacks proficiency with them is like a developer who knows many techniques but not when to apply them. Studying different software patterns and having a thorough understanding of when to apply them allows you to pick the appropriate tool for the job. The right approach makes the difference between solving a problem as an uphill battle and solving it gracefully.

When it comes to dealing with the over-rigidity of rules, a helpful shift in mindset can be to think of rules as "guidelines" rather than strict commands. A guideline points to the spirit of an approach while also acknowledging that there are times to deviate. It's a subtle difference but provides a potent framework for thinking about rules.

Choosing to apply or break a guideline should be deliberate. Simply saying, "Well, this isn't a rule so much as a guideline," is not justification in and of itself; this mentality misses the point of the exercise of adopting the guideline-versus-rule mentality.

Going back to the DRY example, an appropriate level of justification involves building a case as to why code duplication is easier to read in a particular instance. Just like in the webpage example at the beginning of this chapter, it would have made more sense to avoid DRY since the system had a set of similarly structured code that already involved some level of duplication. Deviating from that existing architecture on a one-off basis for the sake of avoiding repetition only served to confuse future contributors.

Using this approach, the term rules can still be reserved for those rare instances when advice does need to be steadfastly followed. Things like "avoid common security vulnerabilities" or "always use version control" for instance, make for exemplary rules in this sense.

Closing the Loop

Much like my mission to incorporate every rule of thumb into all code I would write, I used to view code reviews as an opportunity to show off my acumen by blindly calling out every rule violation on others' code as well. In due time, I became an enforcer of stringent regulations that were of little consequence, and I'm sure this pedantry wasn't helping me win any popularity contests with my teammates.

The real problem was that I wasn't even considering how valid or applicable my feedback was. I wasn't as concerned about providing *good* feedback so much as I was about offering *a lot* of feedback.

In both writing and code reviews, I got caught up in applying the rules for the sake of rules. In doing so, I lost the forest for the trees. What I grew to realize is that most problems live in the gray area of trade-offs where someone needs to worry less about finding a bulletproof solution and more about making the best decision with the information that they have.

Forging your path:

- What types of rules do you consider to be strict rules versus guidelines?

- Has there ever been a time when you've broken a coding rule that you should have followed? What about vice versa: have you ever followed a rule's recommendation but ended up regretting it?

- Take inventory of your own rules toolkit. Which do you frequently reach for? Are there rules that you've heard of but don't yet know what they mean?

The Fear of Failure Pitfall

Accomplishing tasks perfectly the first time isn't nearly as important as practicing and learning from experience.

I had just received the single best piece of advice in my career, and all I did was sit there in confusion.

"You want me to *fail*?!" I repeated back to my manager.

"That's right," he responded. "Well, not fail spectacularly. But at least be comfortable failing within a reasonable threshold."

My performance review had taken a turn that I never would've expected. I thought that out of anyone, my manager would be the one person who *wouldn't* want me to fail. As I would come to learn, he was right.

When I was first learning to write software, I would jump headfirst into playing with new technologies (see The Shiny New Object Pitfall). In due time I discovered those shiny objects came with drawbacks and rejected proposals.

I found myself swinging to the opposite end of the pendulum. Features at work instead became cautiously approached endeavors. When torn between experimenting with a new approach to a problem versus a known method for accomplishing the same task, I would opt for the known version. It didn't matter if there were any potential improvements that the alternative may have offered. I wasn't able to build on my breadth of knowledge because I never went outside of my comfort zone—and my manager recognized this in me. Although it would've been easier for him to just let me maintain the status quo, he pushed me to help reach my potential.

The Perils of Fearing Failure

Failure is often viewed as an outcome to be avoided at all costs. It's common to prematurely analyze results and categorize them as failures. Did a complex feature get delivered but still have an edge case bug? Failure. Did adding a new call-to-action button on the orders page result in the same number of sales instead of the expected 10 percent increase? Failure. Did a developer submit a code review that resulted in a myriad of feedback? Failure.

Each of these examples illustrates the types of problems that developers bend over backward to avoid. Failure comes in many flavors—most of which aren't really "failure" in the true sense of the word. Within this book, failure is defined as the inability to ever achieve a successful outcome.

It's seldom the case that a perceived failure is going to run a business into the ground. Edge case bugs, evolving user flows, and a fledgling developer improving their skills are *not* failures. On the contrary, more often than not, there is ample opportunity to turn failure into success.

In their defense, who can blame developers for getting wound up when things don't go according to plan? After all, nobody *wants* to be wrong. At the same time, jumping right to success or failure as the only immediate outcome is a foolish mindset that prevents you from reaching your full potential in the long term. The developer who places failure on a pedestal will never be able to grow their skills in the same way as someone who has a healthy relationship with risk and is okay with making mistakes along the way.

Experience Is the Best Teacher

When it comes to gaining proficiency in a task, hands-on experience is king.

Consider two scenarios for a developer who is tasked with coding new marketing pages for a website. Each of these pages has images and different layouts to attract people to the site's product. Developer A is dead-set on getting *everything* right before the page is considered done. They go through painstaking preview after preview to make sure the page is an exact replica of the designs, triple-checking that each layout works precisely the same across each browser. It takes the developer two weeks to go through this process. The same pattern occurs for other pages, each time taking two weeks to deliver an additional page.

Developer B is less concerned with having every detail on the page accounted for and more concerned with having it be "good enough." They accept that the layout doesn't match the design pixel-for-pixel in some areas and that different browsers render the page slightly differently. As a result of accepting these blemishes, Developer B manages to submit their code for review in two days. They receive more comments and feedback than Developer A, but each time, Developer B takes this feedback and incorporates it into the next page they work on.

Fast forward one month later. What does the experience level look like for each developer? Developer A has worked on two pages by this point, whereas Developer B has at least ten under their belt. Going forward, Developer B is also better equipped to handle building more pages. They are not only familiar with the code that's needed for new pages, but their skillset is more robust when it comes to handling change. Their experience has weathered them, and they are suited to handle a variety of scenarios with requirements.

Another point that's easy to overlook is that in our scenario, Developer B has ten iterations of feedback during code review from other developers on the team. A bittersweet aspect of programming is that it's both complicated and subjective enough that, despite taking all the time in the world to prepare a code change, someone can always come up with some sort of feedback when presented with code to review.

Sure, Developer B did make more mistakes on the first few pages they worked on, but these web pages were relatively low stakes to begin with. (Of course, this example is one of the vast majority of applications outside the realm of navigation systems for helicopters or a system that controls an electrical grid. It goes without saying that in those mission-critical situations, robust code is non-negotiable.)

Ideally in any scenario, a developer should have the freedom to fail. Good teams are structured to let developers learn from missteps while avoiding dire consequences. In this example, it's okay for a marketing page to have items that are misaligned, but the team should have safeguards in place to help developers avoid mistakes that would take down the entire website.

It can seem like Developer B cares about quality and value less than Developer A, but the truth is quite the opposite. Being able to distill what is authentically useful to the marketing team and deliver on that is an indispensable skill. Quality doesn't lie in being meticulous about every detail; it lies in making sure the imperative things are done right. Software flaws are inevitable—striving for perfection isn't nearly as useful as figuring out what the proper solution is for the task at hand.

It's Not Over Until It's Over

One huge issue with the notion of failure is how quickly teams resign to an outcome when pressing a little bit further could change the framing entirely. Let's revisit the examples at the beginning of this chapter:

Did a complex feature get delivered but still have an edge case bug? Failure. Did adding a new call-to-action button on the orders page result in the same number of sales instead of the expected 10 percent increase? Failure. Did a developer submit a code review that resulted in a myriad of feedback? Failure.

Remember, none of these represent failure when defined as, "The inability to ever achieve a successful outcome." Failure isn't having a bug with an edge case. If the bug is important, it can be patched up the next time the application is updated. The team may even be fine with making a conscious effort to never fix the bug if the circumstances around reproducing it are rare enough.

When introducing that call-to-action button results in no impact on sales, it is an opportunity for further insight. Were users confused by the button? Was its offer not compelling enough? Each time new functionality is introduced and measured in an application, it provides a chance to increase a team's understanding of how users interact with the application. Having a grasp on these interactions helps better inform decisions in the future.

The developer with lots of feedback on their code reviews shouldn't be scrutinized for receiving said feedback. Consider the alternative: if perfection is expected during code reviews, the only time a developer would feel safe submitting code for review is if they had done exhaustive preparation to make everything spotless. This mentality is absurd because it would make the goal of code review to receive no feedback at all.

While it's beneficial to avoid submitting mediocre code for review, there's a fine line between an honest attempt and indulging in perfectionism. Part of every team's responsibility is to foster growth for its engineers, and making mistakes that others can help correct is a key part of the growth process.

Ultimately, It can be hard to avoid conflating the unexpected with failure. Delivering bug-free features that meet metrics every time would be nice if it weren't so unrealistic. Unexpected outcomes test a team's resilience in the face of new information. Rather than proclaiming, "We're done!" more teams should be asking, "What next?"

Closing the Loop

To fail is to grow. Think about the time in your life when you experienced the most growth. Did it result from everything coming easy to you, or did it come after a series of setbacks?

A good manager is someone who knows you well. An exceptional manager is someone who knows you better than you know yourself. My manager's advice was so impactful because it wasn't just a rehash of things I had done well; he knew what I needed to hear in order to actually grow.

It pains me to admit that in the past, my ego has gotten in my way and prevented me from growing at a much faster pace because I was afraid of making a mistake. Paradoxically, by striving for great code, I often found myself working in a vacuum for too long and ended up with a worse skillset in the long run. There *are* some rare developers who can read pages upon pages of documentation and synthesize its concepts into code on their first go around; I'm not one of those rare developers.

Over time, I've found that the best way for me to learn is to play around with new concepts by writing some code and quickly getting feedback. In this scenario, I find myself in a place where I'm removed from my natural inclination to want to protect my code, and I'm able to view feedback as a way to collaboratively come up with the best code possible.

Forging your path:

- Where is the line for what constitutes "good enough" software? How does it vary from domain to domain?

- When is a solution a true failure and when is it a perceived failure?

- Consider the application you're currently working on. What parts of it require solutions that need to be right the first time around? What areas have more flexibility for experimentation?

PART II: WRITING CODE

The Development
by Hope Pitfall

Cutting corners is tempting with complex problems, but those are the problems that require the most diligence.

I was altogether exhausted as our team closed in upon the end of our sprint. After tirelessly working on a new "project wizard" feature for the previous quarter, I accepted that my fate now belonged in the hands of the code gods. Against my better judgment, I was ready to knowingly commit a software sin: deploying a change that I knew would fail.

It's not that I enjoy pushing out fragile code... nobody does. In fact, I consider myself to be overcautious about my code on most features when I'm under a normal workload (see The Fear of Failure Pitfall). But, in this case, my lack of planning and circumstances caught up with me.

Rather than setting up automated tests, I focused on writing the happy path feature and testing everything by hand. The project wizard involved tediously filling out information over the course of a five-step flow. Testing those last two steps didn't get nearly as much attention as the first three steps.

To make matters worse, there were lots of combinations of logic to be tested. Our QA team would stumble upon hard-to-reproduce scenarios that would only crop up with very particular combinations of inputs. I knew that it wasn't a matter of *if* a user would encounter an error, but *when*.

Yet I still fixed the bare minimum of bugs found by our QA team and pushed the feature out. Why? You could say it was due to carelessness, but I don't think that quite pins it down.

Ok, part of it *was* carelessness, but it was also that I was worn out after working on such a big feature for so long. I had also, in retrospect, given in to the sheer enormity of the task. I knew that even a feature that was rough around the edges could be shoehorned in before code freeze and could later be patched up with subsequent bug fixes.

Ultimately, I was more beholden to delivering a certain threshold of story points within a sprint to maintain the illusion of progress. Even inferior code counted towards maintaining the team's velocity and kept management appeased. In short, I betrayed the long-term benefit in order to avoid rocking the boat in the short term.

The Perils of Development by Hope

There are times when working on a feature goes swimmingly. The problem is obvious, the solution is obvious, and the steps to get there are obvious. Testing can be done confidently, and there aren't many edge cases to worry about. Stumbling upon such problems is like wandering into a zen rock garden; knowing the path is calm is enough to induce inner peace.

On the other end of the spectrum are the unruly bugs and features that are hardest to wrestle with. Those bugs that take fifteen steps to reproduce or the ones that seemingly occur randomly. The feature that interacts with a subsystem so unstable that it's an absolute minefield.

When put under pressure in these strenuous circumstances, it can be easy to succumb to development by hope. Hope that pushing forward with a code change will fix a problem, but lacking any evidence to back it up because it's hard to test. Or hope that a limited understanding of what's being requested is enough, despite knowing that a problem is more complicated than how it appears on the surface. At its core, The

Development by Hope Pitfall means turning a blind eye to the diligence involved in addressing complex functionality.

Although software development can be convoluted, hope isn't the answer. Throwing code into a product without a material understanding of what it does is a recipe for disaster. Yet this practice happens all the time—be it from tight deadlines, laziness, or fear of complexity. Code written in such a way will need to be reworked again later, and it's easier to write clean code the first time than it is to retrofit a poorly implemented solution.

Becoming Fearless

The solution to development by hope is the code equivalent of, "When the going gets tough, the tough get going." Working on fierce bugs and nebulous features is less about writing code and more about the approach.

Consider using this question to guide your approach on such tasks: *What do I need to be confident about shipping this update?* This forethought is critical because without context and confidence, fear can kick in. Fear of the hard-to-reproduce bug rearing its head again after a fix goes out. Fear of not developing the right feature because what was being requested barely made sense in the first place.

What's needed to be confident about an intermittent bug fix? Being able to reproduce it consistently. Without that, there's no knowing that the bug has been fixed. Given this goal, the tactics to employ begin to crystalize: add logging to gather more information, narrow down as many variables as possible, and write tests. Embracing short feedback loops helps immensely too. If a bug takes fifteen steps to reproduce, see if it can be distilled down to fewer steps or automated altogether. By optimizing for confidence, the odds of success are greatly shifted in your favor.

What's needed to be confident about implementing an unclear feature request? Being able to explain the feature in plain English. A better understanding of a request could come from diagrams or a better description in a ticket.

A simple yet overlooked step is to have a conversation with whoever is requesting the new feature before writing any code at all. When given unclear directions, unabashedly asking questions over a quick video chat saves an extraordinary amount of time compared to making a best guess that inevitably needs to be fixed later. Sometimes this takes patience, especially when it can be hard to have to wait for the feature requestor's availability to have this conversation. Other times this takes setting aside one's ego and not being afraid to ask questions that appear to be uninformed.

Regardless of the obstacles, gaining understanding of a feature is well worth its modest cost. Going down the path of writing code for a feature you don't understand is a waste of your time. In the same way a bug can't be knowingly solved until it can be reproduced, a feature can't truly be implemented without comprehending it. *If you can't explain it, you can't build it.*

Closing the Loop

Remember the project wizard example? An alarm should have gone off in my head as soon as I saw that it was tough to reproduce the latter steps of the process. At that point, I should have looked to create a full-fledged integration test that could walk through the entire survey flow effortlessly.

Sure, it would have taken more work to set up in the first place, but that work would have paid dividends for every developer on the team going forward and allowed me to ship the feature confidently.

I don't develop by hope anymore. Instead, I tackle pain points head-on. Software shouldn't be guesswork. Although it can feel faster to take a shortcut, it can just as easily result in rework. Writing code confidently embraces writing it correctly the first time.

Forging your path:

- What types of bugs or features do you fear working on in your current system? What about them invokes fear?

- Are there developers you work with who appear fearless? How do their workflows inspire confidence when working on tough issues?

- How often do features you work on require follow-up conversations with your stakeholders that could have been avoided if more requirements had been gathered upfront?

The Shiny New Objects Pitfall

Experimentation with new technologies is crucial to becoming a well-rounded developer, but adding them to a system should be done judiciously.

I was ecstatic to have found a new WYSIWYG rich text editor to use on our website. Why, you ask, did I care so much about a mundane feature? Because the tool we had been using at the time was one of the biggest pain points in my daily workflow. There was constant upkeep that it needed to function on older browsers we supported, and anyone who has ever had to support a rich text editor knows that the way it stores information can quickly become a nightmare.

Our company had monthly hackathons where we carved out time to work on experimental features. I was filled with hope and excitement when I was able to get a proof-of-concept with the new editor working in just one day.

To say I gave this new library the benefit of the doubt is an understatement. I was smitten with it, absolutely ready to buy into its promise of being a silver bullet to all of our problems with the old editor. Even though I hadn't pushed its limits beyond an initial implementation, its installation was such an easy process that I couldn't fathom its full path to implementation as being one with many obstacles.

My product manager asked me about licensing for the new library. I avidly told him about the one-time $400 payment and how great of a deal that was for a company our size. Rightly, the product manager was skeptical. "If it only costs $400 once rather than a subscription, that means that their team isn't incentivized to give ongoing support to the library in the long term," he pointed out.

Of course, he was right—and his remark has stayed with me years later.

The Perils of Shiny New Objects

Software development is a novel discipline in that there is constantly something new around the corner. New frameworks, new languages, new tools—all begging to be tried.

The Shiny New Objects Pitfall occurs when the line gets blurred between being interested in something new and going all in on its implementation. These shiny objects can seduce teams with their promises of performance and productivity, not to mention that newer projects tend to be updated more frequently, which carries an inherent degree of excitement.

But newer doesn't mean better. Troubleshooting issues with a newly invented programming language can be difficult. There are fewer resources available where people may have stumbled upon analogous issues—making it harder to find the exact solution you're looking for when you come across a problem. Additionally, new tools that are actively maintained by one or two contributors run the risk of being abandoned.

Something funny happens when evaluating different frameworks and libraries to solve problems. There is this inherent benefit of the doubt that's bestowed upon new solutions. We tell ourselves that the new payroll software is going to fix all of the problems from today's system, the new UI framework is going to address all of the pain points of the existing framework, the list goes on...

The fact of the matter is that while shortcomings may very well be apparent in our existing systems, it's much harder to anticipate the problems that can arise when we start integrating new solutions. Sure, the proposed UI framework can handle mobile viewports better, but will it be able to handle that immaculate new layout a few months down the road? Maybe,

maybe not. Maybe the old framework would've handled it better. Regardless, it's impossible to say definitively. The important takeaway is we should reject the assumption that our new, shiny solution will fix *everything*.

Evaluating different solutions involves taking a hard look at the pros and cons of each of them. If there aren't any cons for a new solution, that's a huge red flag in the evaluation process. Remember, if something sounds too good to be true, it probably is.

Hobbies versus Professions

Most people only have one profession and many hobbies. There is a decision that we subconsciously make about what should be a hobby in our lives versus what should be a profession.

Usually, this choice is evident. If you went to school for teaching but picked up a tennis racket for the first time, for example, you're unlikely to be convinced that jumping into a career change as a professional tennis player would serve you in the long run. Instead, it's probably best if you leave it as a hobby to be explored and enjoyed in their free time. Other times, the choice is more subtle: say you love working as an accountant, but acting has always been your passion. You may eventually decide that it's time to switch paths. By making a career change to become an actor, you've upgraded acting from a hobby to a profession.

Neither of these decisions is right or wrong. These examples merely serve to illustrate that a person can be passionate about many things at once. Since it has such profound implications, choosing a profession is a decision that requires rigor.

When it comes to the hobby versus profession analogy in software development, hobbies represent all of the shiny objects like new frameworks, languages, and tools, whereas professions are the technologies that are employed in a project at work.

A software developer who suffers from shiny object syndrome attempts to make every hobby a profession. Every professional project they work on must incorporate the latest flavor of the week—whether that be a new language or library—regardless of how applicable it is to the project. This plays out to their own detriment. That expense reporting software *doesn't* actually need to incorporate a fancy new animation library. Adding animations unnecessarily means that these types of features need to be supported by the engineering and design teams down the road.

Make no mistake, hobbies are fantastic. They expand variety in our lives, offering sources of inspiration and passion. The teacher who plays tennis is more well-rounded, be it from the exercise, socialization, or insights into techniques that the sport provides. Likewise, shiny new objects expand a developer's skillset and teach them new techniques and ways to think about problems.

The key is not to be allured by attempting to incorporate them into *every* work project. Instead, aligning the right technology with the right situation sets you and your team up for success.

Closing the Loop

Looking back, I'm convinced that it would have been a disaster if we had gone with that new WYSIWYG editor. I had been too caught up in what a great deal the product would've been for us to consider the ramifications and potential liabilities of relying on an outdated library.

Engineering-wise, I'm certain that we would have traded one set of implementation problems for another. Those types of deeply nested challenges don't surface until you're knee-deep into its performance within a larger user base.

But, again, shiny objects aren't all bad. In fact, they are crucial to learning and growing. A software developer who has mastered shiny object syndrome can accurately gauge how much freedom they have for experimentation on any given project. If a project has a clear-cut objective and deadline, then it's a good match for tried-and-true technologies. If a project is a greenfield application with novel requirements, leaning into more creative ideas and using a shiny new tool can be just what the team needs. Certain projects can afford more freedom to play around with experimental features.

There's a balance to be struck somewhere between relying upon proven technologies to provide the foundational base for a project and acknowledging opportunities to enhance an application with shiny objects.

Forging your path:

- What are the chances that every new tool you come across solves a problem entirely? What types of problems lend themselves well to having existing solutions and what types require custom code?

- When it comes to using different languages and frameworks, when is it better to be a jack of all trades versus a master of one?

- Have you ever employed a shiny tool for a project and then regretted it? What about the opposite, have you ever utilized a known framework when a newer tool would have done a better job?

The Wheel Reinvention Pitfall

Writing custom code for a solved
problem is more error-prone and less efficient
than using an existing solution.

"So how did you get that data? Did you use an API?" my interviewer asked.

"Nope, I scraped it all myself," I proudly answered.

"Scraping as in writing custom logic to parse the webpage? That's a bit tedious, isn't it?" he prodded.

I wasn't catching on. "Yes, but I made it work with the custom code that I wrote instead!" I told him. I may as well have also tacked on that the project wouldn't have succeeded without my brilliance as long as I was being so oblivious and wrong. He nodded and continued to the next question without belaboring my misstep. (This interview was one of my first interviews right out of college, so I shouldn't be so hard on myself for what I now consider a cringe-worthy response.)

It's one thing to realize you're wrong in the moment, but it's another altogether to be so blissfully ignorant that you actually *brag* about doing something terribly. It's like if you took your car to a mechanic to have your brake fluid reservoir fixed and they bragged that they saved you time and money by reusing a soda bottle instead of the actual part.

I suppose I was enamored with my own ingenuity and naive enough to think that there was no better way of getting data from a website than scraping it myself. Being so new to software development placed my blinders on quite snuggly. I didn't pause to stop and think about how or why an existing solution may be better.

The Perils of Reinventing the Wheel

Software development is an act of creation. What's interesting about coding is that pieces can and should be reused. To build something new, you can leverage building blocks that others have already created to solve common problems. Modern software applications are nothing short of a miracle in the sense that they are a fusion of different libraries that people all around the world have worked on.

Developers fall into the Wheel Reinvention Pitfall when they opt for brand new code instead of using a well-established version that already exists to do the job. Particularly at the beginning of a software developer's career, there's a temptation to rewrite functionality from scratch—be it for a user login flow, an analytics tool, or even something as basic as an array-sorting algorithm.

It's important to note that this pitfall applies to these types of *generalized* problems. That is to say, the problems that are so common that solving them in one application looks the same as solving them in another.

Maybe someone should in fact implement a custom sorting algorithm on a very specific use case where they have a million records of data and some quality about that data makes it ripe for optimization. If the problem is genuinely unique, reinventing the wheel could be warranted.

But having such perfect alignment is almost never the case. Implementing a custom sorting algorithm on an array doesn't make sense on a collection of a hundred items. Nearly every programming language already has a standard library with a battle-tested sorting algorithm that will do an excellent job of handling this task, and the difference in performance on such a small dataset is negligible.

This temptation toward rolling your own implementation can stem from multiple causes, some more justified than others.

There might be a library that solves a precise problem, but not the exact problem your team is tackling. Or maybe existing libraries have issues that your team is looking to overcome.

While writing a new implementation can seem like a credible idea, there are plenty of issues that come with it. Writing a custom solution means that the developers on the team are responsible for every facet of the solution, from the code's functionality to testing to performance to documentation—an entire suite of obligations that ensures ongoing maintenance incurred by the team.

On the other hand, existing tools help address these sorts of responsibilities and are often quick to implement. It's never been easier to leverage popular open-source libraries, whose integration into an application can be added with a mere few lines of code. These libraries have contributors who specialize in solving, optimizing, documenting, and maintaining a solution that's narrowly scoped to the library's domain.

In our earlier example of adding a user login flow, an authentication library can utilize best practices. A popular library is going to cover standard conventions like making sure passwords are hashed and appropriately expiring a user's session. A team *could* write a custom version of a login flow, but they'd have to solve many of these same sorts of problems on their own anyway. Not only that but there is a high likelihood of making a mistake in the process.

Finding the Wheel

With experience comes a knack for deciding when to reach for an established solution and when to write code anew. By finding the right balance, you can greatly amplify your productivity.

The first step toward not writing code that already exists is awareness that an alternative option is available. That alternative may live elsewhere in the project, in the standard library for the language, or in a third-party library.

One compelling approach is to default to using the existing solution unless there is a strong reason not to. Justification for *not* using an existing solution could be:

- There is existing code written specifically for a different domain that shouldn't apply to the current use case. The process of trying to buy a house is much different than buying a car. It would be a mistake to try to reuse all of the logic from a home-buying software application on a car-buying application.
- An open-source library has a license that is too restrictive for the project.
- A free-to-use API is no longer actively maintained and will likely have compatibility issues going forward.

Another litmus test for whether a problem is a worthwhile candidate for using an existing solution is if that problem is *not* a core value being offered by the product. Suppose a team is working on tax-filing software. The main focus of the product is to make taxes easier to file, which means that the team shouldn't be focusing on trying to write their own custom login flow when it just needs to be the same as every other login flow out there. If their sales team is going around telling prospects, "Our tax application has a login flow that's going to blow you away!" as a major selling point of the software, then something has gone horribly awry. The most valuable proprietary code that they write is related to the intricacies of tax rules, as that is at the core of their offering.

Deciding to build a new solution for a problem can be a costly endeavor for a team, as the product development lifecycle can take many people and lots of time. It's something that should be done consciously and with a definitive rationale.

Closing the Loop

I wish I could say that interview long ago was the only time in my career I had fallen prey to reinventing the wheel, but the truth is that this trap has ensnared me in many forms over the years. Sometimes it's a relatively small piece of an application like that scraping logic. Other times it's as large as an entire messaging and notifications subsystem. Ironically in almost every case, using an existing product/tool/library would have not only been easier, but it would have provided better quality.

At the crux of product development lie the many factors to consider before reinventing the wheel. How much utility does the feature provide? Are there options that build upon existing solutions to address the majority of the problem being solved? By weighing these factors, reinventing the wheel is saved for unique circumstances where it may be applicable rather than an eagerly embraced crutch.

Forging your path:

- When is a time that you wish you would have used an existing solution over building a custom solution, and vice versa?

- Take stock of the languages you work most with. How familiar are you with their standard libraries? By acquainting yourself with these libraries, you're far less likely to reinvent the wheel when manipulating low-level data structures within a language.

- Are there areas of applications where you find yourself most prone to reinventing the wheel?

- Have you ever had trouble convincing a team member to use an existing library or framework? What was their hesitation around this decision?

The Writer's Block Pitfall

An action bias is crucial for writing software.

When I was new to programming, I rarely had writer's block since I would jump headfirst into problems without inhibition. I didn't know where I was going half the time, but that didn't stop me from throwing random code at the wall to see what would stick.

Fast forward to the present. Now writer's block is not really an issue because I have a toolkit of strategies to deal with various types of problems that may arise. I'm able to quickly identify when I'm coming down with writer's block and employ one of those strategies to ward it off.

It was during that middle stage a few years into my career when writer's block hit the hardest, and the specific combination of a difficult problem along with deficits in my skillset seemed to set it off.

When I was an "I know enough to be dangerous" type of mid-level developer and given an expansive task, I froze. I was experienced enough to know that large problems needed time and attention to architect a proper solution but became paralyzed at the thought of writing anything down. I knew that my first draft of code wouldn't cover the edge cases that lurked around the corner. Although I was in familiar territory, I found myself lost.

The Perils of Writer's Block

Writer's block needs no formal introduction. Just like in the non-engineering world, it can be an infuriating experience to stare at a blank page, knowing that *something* needs to be written but not knowing exactly where to start.

How do we arrive at this impasse? Sometimes vague requirements induce a state of confusion. Sometimes a problem is overwhelmingly huge. Sometimes it's just the human condition after sleep deprivation or neglecting food. Whatever the cause, the resulting feeling of stuckness has a universal result—your progress screeches to a halt.

Unblocking Yourself

What makes or breaks writer's block as a pitfall is how you handle it. If you're only stuck for a few minutes and can snap out of your stasis, it's not too harmful. When writer's block becomes a true pitfall is when it's used as an excuse to avoid a problem altogether. Putting off a task for hours or days only serves to make the problem seem even more daunting. When writer's block hits, the hardest part of approaching the problem can be finding the motivation to get started; having a handful of strategies for getting back on track at the ready can make all the difference.

Embrace the Rough Draft

Rather than trying to be pristine the first time around, embracing a rough draft means that code can be cleaned up later on. This frees the author from worry—whether that be the worry of finding that ideal encapsulation, the worry of code duplication, or the worry of conjuring that perfect variable name. Just getting something on the page that works is the sole focus, however ugly it is in its initial form.

This approach works well with mysterious problems that trigger "analysis paralysis." When problems can be solved in multiple ways, it's helpful to have a clear process for sussing out high-level abstractions.

It bears mentioning that this strategy is not an excuse to submit low-quality code for review. It's not only fine but encouraged to post experimental code for a "draft" review in order to solicit feedback from others quickly. When it comes to submitting a formal code review, though, rough drafts no longer cut it, and the quality of the code should measure up to the bar set by the team.

Pair Programming

Nothing jumpstarts the code-writing process like having someone else present to help hone your full attention on a problem. Of the many benefits of pair programming, warding off writer's block is an under-acknowledged topic.

When a problem has multiple controversial approaches, a partner can help weigh the pros and cons of different approaches and assist in deciding on a path forward in real time. Without pairing, the feedback loop can take hours or even days as the code makes its way through review.

Pairing is particularly effective for problems that require juggling many mental contexts. Between the driver and the navigator, the driver handles the minutiae of code being written from syntax typos to runtime errors to environment issues that may arise. The navigator can, in turn, dedicate their attention to different facets of code architecture. The complementary nature of these roles ensures you maintain momentum. Done thoughtfully, and writing code in this manner is smoother than if either person had done it on their own.

Outline the Code

Trying to find the proper encapsulation for a problem can be difficult. Sometimes it's not until after writing a full-fledged solution that things like high coupling and leaky abstractions become apparent.

An interesting way to take a stab at figuring out how to organize code for a problem is to first write some plain English steps in an attempt to outline it. For example, if the task at hand is to import data from a third-party system, the steps of an outline could look something like this:

1. Authenticate with the third-party API.

2. Make an API request to fetch the data.

3. Read each row of data. Within each row, validate the information and create a corresponding record in the system.

4. Notify the appropriate users how many records were created and how many errors were encountered during the process.

Writing an outline provides hints as to which areas of code should be grouped together and which should stay separate.

In this example, writing authentication logic and coupling it with the completely different step of notifying users about the result of the import leads you down the path of writing what is infamously known as spaghetti code. As soon as multiple steps of the outline start bleeding together in the code, it's probably time to revisit if the chosen abstractions are logical.

A related approach can be to use test-driven development as a way to outline how code should work. A capable testing framework supports syntax that can closely mirror requirements, so writing test descriptions like "when an item has changed price since it was first added to the user's cart, display an informative message" allows a developer to explicitly articulate the expected behavior. Even without implementing any of the test code, having the descriptions written down can make a big feature feel like less of a mental burden.

Take a Break

If you're hesitating to tackle a problem because it feels intimidating, taking a break might sound counterintuitive... but it works and it's hard to argue with results. Going for a quick walk or stepping away from your desk for a few minutes can aid in dealing with writer's block because when you're drained, you often need to relax to give your mind some breathing room. A break allows you to put your best foot forward when picking a task back up.

Notably, it's hard to resist feeling guilty about taking a break. But the benefits of a break become obvious with a hypothetical. When entrusting someone to complete a sophisticated task correctly, is it better that they have a well-rested mind or that they start the task immediately, regardless of how mentally drained they may be from their previous task? Tracking hours and rigid schedules are the levers that many people reach for in order to produce results when actually they should be focusing on getting the foundational elements right that will set them up for success.

Closing the Loop

In the thick of my stuckness, I was too ashamed to admit that writer's block was a problem for me. It wasn't until I turned the corner that I was comfortable talking about it with my colleagues. Unsurprisingly, many of them had the same struggles I did, and I wish I had been vulnerable earlier on so that we could've helped one another. Letting go of my ego would have paved the way for growth much sooner.

Forging your path:

- In what types of situations do you find yourself struggling with writer's block the most?

- What strategies do you reach for when it's time to resume work after a break?

- How frequently do you incorporate breaks into your day?

The Multitasking Pitfall

Trying to do many tasks at once is particularly detrimental for coding.

I opened my laptop for the day, excited to have a fresh start with no meetings. I had one sole task to code up for the day and was looking forward to immersing myself in it. I started up my development environment and began diligently coding away.

Ping! A teammate reached out to me on chat to ask me about something. I instantly replied to him and had a few more back-and-forth messages over the next fifteen minutes. I resumed my original task.

Bloop! An email arrived from HR reminding me that quarterly self-evaluations were due in a week. I told myself it wouldn't take long and may as well finish it while it was top of mind. Half an hour later, I went back to coding.

More people arrived at the office. *Ping! Bloop! Ping!* Visitors at my desk. I should have realized that things never get *less* busy as the day progresses.

The end of the workday rolled around, and I wondered where the time had gone. I didn't have much to show for the day other than what remained in my chat logs and sent email folder. At that point I realized that being busy is not the same thing as being productive.

The Perils of Multitasking

Multitasking has the illusion of value but in reality leaves its practitioner in a worse state than where they began. It may seem like someone who is constantly going from emails to chat messages to writing code is maximizing their efficiency, but taking a closer look at multitasking reveals a different picture.

The Multitasking Pitfall occurs when you miscalculate what you're able to (or *should*) take on at once. Much of the time it's okay to switch back and forth between tasks that don't require as much involvement. Organizing emails while waiting for an application to finish compiling, for example, can of course be done at the same time. However, as soon as even one highly demanding task is introduced, multitasking fails miserably.

Anyone who's ever written code knows it's an exercise that requires utmost concentration. Adding one seemingly small task like answering chat messages while coding takes an instant toll on the quality of each. Sure it can be done, yet interruptions of any size affect not only the quality of the tasks but ironically inhibit the speed of completing them as well.

By multitasking, you forgo the precious ability to focus and legitimately dig into problems. Focus is vital if you're trying to build a mental model around a problem and become familiar with the different pieces of code at play within a system. In addition, jumping from task to task in quick succession creates context switches, whereby it takes time to adjust to a new task after being mentally engulfed in another. The tax incurred by context switching is particularly high in programming. It takes time to rebuild mental models after a context switch shatters them.

Long Live Single-Tasking

Single-tasking is the unsung hero of productivity—you'd never hear someone say, "Wow, look at them! They're such an amazing single-tasker!" Well, that's partly because that's a strange thing to say, but I can tell you that I've absolutely heard about people being praised for always being the busiest-looking person in the office.

But the best developers I've worked with weren't the people who always looked frantically busy. Those who could produce the best code day in and day out were those who had mastered single-tasking. By leveraging the following techniques, you'll be able to take your single-tasking skillset to a new level:

Batching

Batching is the practice of doing similar tasks together. For example, this could look like going through all pending code reviews in succession or setting aside half an hour for email correspondence at the end of the day.

Impromptu requests negatively affect your ability to batch tasks. To succeed at batching, you must come to terms with the fact that the vast majority of last-minute requests can wait until later in the day or week.

The reason batching is so powerful is that it requires an active commitment from its practitioner. Once you've committed to performing a series of tasks, you are freed from distraction. Rather than incurring context-switching costs, related tasks may even gain momentum. It's easier to answer the fifth email in a row once you're in the rhythm of writing than it is trying to sporadically compose messages while also doing other work.

Scheduling

An effective schedule is one that leans into how its owner operates. If you're a night owl, you may prefer to take on mentally intensive tasks like performing code reviews in the afternoon or evening when you tend to experience your highest level of cognition. On the other hand, if you're a morning lark, you may do just the opposite by preferring to review code in the morning, saving lightweight tasks like answering emails for later in the day. Knowing when you do your best work allows you to leverage your schedule strategically.

Let's name the elephant in the room here: our schedules aren't always within our complete control. Team obligations do pop up, and the reality is that some days will be more meeting-packed than others. One way to prioritize how a day is organized is to explicitly schedule time for personal tasks. It's reasonable to carve out time by blocking off one's calendar for everything from writing code to filing expense reports. Formalizing duties with a timeblock prevents the day from being overtaken.

Deep Work

Deep work is the polar opposite of multitasking. There's no hopping from one task to another; instead you're concentrating on a single task at a time. Deep work requires solid chunks of time, preferably an hour or more, without interruption.

The influence deep work has on the process of writing code cannot be overstated. Creating software involves internalizing many factors, from design patterns to tests to edge cases.

Working with code for a long period of time allows a developer to become absorbed with the subject matter and promotes clarity. In fact, a developer immersed in deep work becomes similar to a chef working in a kitchen with which they are intimately familiar. The chef knows exactly where to reach for different ingredients and utensils, leveraging this familiarity to consistently produce well-crafted dishes. By working in a problem space long enough, a developer knows exactly where to reach in different areas of code, internalizes how to reproduce relevant scenarios, and accumulates a sense of the domain at large to consistently produce well-crafted software.

Doing this deep work can lead to *flow state*—the experience where someone operates at a high level effortlessly. Flow state is being "in the zone;" for coding it's when a developer knows where they're going and how to get there. Keystrokes automatically fly in service of a honed destination, and in some ways it feels like the code is writing itself.

Flow state can't just be turned on with the flip of a switch. However, it can be nourished by embracing deep work. The undisturbed, focused environment that's created bolsters conditions that are conducive to this mode. The more deep work is practiced, the more patterns emerge with finding flow state.

Closing the Loop

My confession is that I used to be a chronic multitasker. Any time an email or instant message came in, I would rush to reply to it so that people knew they could count on me. In doing so, I hoped to build a reputation as being someone who was trustworthy and efficient. This attitude came at the price of my concentration. Instead of doing one task well, I did many tasks poorly and took longer than I needed to.

The good news is that there's hope for my fellow multitaskers. As soon as I stopped trying to do everything at once and instead focused on doing one thing at a time (and doing it well), I noticed a huge boost in my productivity and quality of work.

Forging your path:

- What types of tasks do you combine today that would be more efficient to be done via batching?

- How many multi-hour blocks of uninterrupted time do you have throughout the week to write code?

- How could your day be structured differently to optimize the tasks you have to do?

The Code Delivery
en Masse Pitfall

Submitting large code changes all at once
helps neither the author nor the team.

Early in my career, I was always nervous when the time came to deliver large features. I recognized that these features are inherently accompanied by potential rewriting and retesting, and my way of coping with this fear and minimizing my pain was to deliver all of my code for review and testing *at once*. I treated my code as something that needed to be kept under wraps until it was ready to share with the team.

Delivering one giant chunk of code obscured my teammates' ability to thoroughly critique my code within a given timeframe. There's an old joke that if you assign a 100-line code change to a developer for review, they'll come back with twenty items for feedback; whereas if you assign a 5,000-line code change, they'll just say, "Looks good," and approve it. Amusing as this may be, the point is that when a code change hits a certain threshold, many people become too overwhelmed to provide quality feedback.

The same thing goes for the quality assurance process. QA teams with strict sprint windows are beholden to limited timeframes for testing. If they ran out of time and things looked "good enough," even subpar code is permitted since nobody wants to hold up a release.

It's worth acknowledging that there is an innate degree of time sensitivity in a sprint. In these cases, "done and imperfect" is an acceptable and common expectation. That is different from imposing time limitations on a project unnecessarily or rushing "because you can," which can result in an unacceptable degree of risk.

In my case, the choice to deliver code all at once wasn't one that I made maliciously. Admittedly, I didn't *like* having lots of feedback or tickets that were rejected in the QA process, but I wasn't setting out to trick anyone. I had convinced myself that minimizing my pain was best for the entire team, but in reality it was what seemed best for me at the moment.

The Perils of Code Delivery en Masse

Every time a developer writes new code, a decision is made around how to group or divide functionality and which parts should be delivered piecemeal. Sometimes this decision is obvious; sometimes it is not. For small tasks, delivering everything at once makes sense. There isn't much efficiency to be gained in delivering a ten-line bug fix in multiple pieces.

The question becomes more interesting with tasks that operate on a larger scale of code. As a pitfall, code delivery en masse is making one sweeping update rather than doing so incrementally in smaller, more manageable pieces.

Consider a feature like writing a user sign-up process for a website. Let's say that this is a multi-page process where a variety of information is input from the user. The engineering team estimates it will take a developer about two weeks to ship this new experience. The product manager is happy because this estimate fits snugly into their team's two-week sprint cycle.

The developer feels that this two-week estimate is tight but attainable. They dutifully plug along throughout the entire sprint, diligently making sure that their code matches the specification laid out by the ticket and that the user interface is a spitting image of what the designer gave them.

Right at the end of the sprint, the developer submits all of the code as one large 2,000-line change. Their fellow developers have some suggestions for changes, but nothing that prevents the change from going in. Since the change does technically allow users to sign up, the QA team approves the change after less rigorous testing because they are near the end of the sprint. Relieved, the developer merges their code, and the sprint's deployment goes out on schedule. The product manager is pleased that the developer was able to finish this project on time and is already looking toward the next set of features.

Product development too often occurs in this manner. Hasty techniques are accepted and even praised if it means a deadline can be met. You may be thinking: *What exactly is the problem here, though? How is this even an example of a pitfall if the developer was able to deliver this feature?*

In short, code delivery en masse trades speed for feedback opportunities. These lost feedback opportunities come in a variety of mediums. Let's reexamine our example through this lens:

- The code reviewers were worse-equipped to spot bugs or call out refactor opportunities given the size of the change and the timing of its submission during the sprint. If a showstopper code issue had been found, the feature may not have made the sprint at all.

- By feeling a time crunch, the QA testers weren't able to thoroughly vet edge cases and thus opted for verifying the happy path flow. By rushing now, the team will undoubtedly spend more time with follow-up bug fixes later—a scenario that is not only bad for developers, but bad for end users too (not to mention the company's bottom line).

- Product team members were shortchanged in terms of feedback opportunities. Even excellent designs can behave differently once someone is able to play around with them on a webpage. Occasionally, a field is cumbersome to fill out or looks out of place on a handful of devices during real-world usage. Delivering a feature all at once right at the end of a sprint forgoes time to provide feedback that only hands-on experience can provide.

Now what if the developer had delivered this code incrementally? In place of a single change for the entire sign up flow, let's say the developer creates multiple smaller, targeted changes—one for each step in the multi-page sign up flow. In doing so, *the developer makes continuous progress toward the goal.* To avoid having partially complete code in a production environment, the developer in this example uses a feature flag to gate access to this new experience. Because these changes are smaller in scope, they are quicker to write and therefore go through the development, code review, and QA process every few days as opposed to once in two weeks.

In taking the incremental approach, the developer has empowered the team with more feedback opportunities:

- Code reviewers can contribute useful feedback when a change is small enough to be easily internalized by the reviewer.

- QA testers can test the sign up flow one page at a time as it is being developed. Since the changes are made available sooner in the sprint, the team has time to dedicate to edge cases, regression testing, and test automation.

- The product team can try out each of the sign up pages as they go into the testing environment throughout the sprint. Feedback on the user experience can be noted immediately and possibly reprioritized into the sprint, thereby avoiding the need to wait until the next sprint (as would have assuredly been the case with the all-or-nothing approach).

When to Deliver Incrementally

Deciding when to deliver a large change is a challenge of its own. Here are a few factors to consider when determining to deliver code piecemeal:

What is the certainty (or uncertainty) of the task? Tasks that trailblaze a new user experience are better candidates for incremental delivery because that approach allows product stakeholders to engage with the new functionality, exploring what works and what doesn't.

Contrast this with a well-known experience in an application. A feature like adding another report when there are other existing reports with the same format, UI for accessing them, etc. is less prone to having much feedback once deployed. If a task follows established patterns in the application and the domain is unambiguous, then the efficiency of delivery en masse could justify the trade-offs.

Is there a real deadline? Deadlines are frequently set arbitrarily. For example, consider a task that's "due" just because a room of developers estimated it could be done within a sprint. That's not a deadline; that's an estimate that's been misapplied as a deadline.

A true deadline is imposed by an external factor. In these cases, delivery en masse could be a reasonable and conscious choice. A team may acknowledge that having a working demo is more important than having every edge case covered, for example, when prepping for an investor meeting.

Are there organizational constraints? Consider external factors—i.e., those outside of your control—before jumping into delivering a feature incrementally. A system can be so massive that it only gets updated a few times a year. Or it might be the case that every change requires a code review, and the reviewers themselves only have limited time available throughout the week.

Regardless of the technicalities of the situation, any large feature should be approached with tact. Just because that massive system has long periods between deployments still may mean you work on components of a feature individually while still testing the functionality in large swaths. Likewise, even if a code reviewer claims they only have time to do one larger code review, the author of the change can organize their approach into smaller commits so they can better explain how the respective steps fit into the larger scheme of what's being proposed.

Can you leverage a bias toward incremental delivery? In general, defaulting to incremental delivery is a safe bet, as it is easier to add more code to a change than it is to break down one big change into smaller units that still work well on their own.

Leaning into patterns like using feature flags can allow a system to gracefully explore new functionality without having releases blocked by partially complete work. By choosing to deliver incrementally, you are forced to consciously determine what the significant steps are, thus giving more scrutiny to the problem itself and setting yourself (and your team) up for greater success.

Closing the Loop

It wasn't until I started being more strategic about large features that I realized how backward my old habits were. As soon as I started to focus on what smaller code changes would be easiest for my teammates to review and test, my code improved by leaps and bounds.

My underlying fear of large deliveries had effectively disappeared. Rather than hiding changes from coworkers, I accepted their help in making my output better.

Previously, there were rare instances where I had taken the wrong approach *entirely*, only making this discovery at the tail-end of the development process. This was always excruciating to experience because it pointed to two things: that so much time and effort was wasted and that said waste could have easily been avoided. Having a quick feedback loop from teammates would have allowed me to course-correct much earlier in the process.

When done thoughtfully, a feature delivered incrementally flows smoothly through the development process—from writing code, to code review, to QA testing, and finally to deployment. One of the greatest services a developer can provide for their team is to make this entire process as seamless as possible.

Forging your path:

- What factors make you inclined to deliver code en masse? Does this happen when you're on a tight deadline? Does it happen when you are working with large features? Is it something else entirely?

- Think about the last time a teammate of yours shipped a large feature successfully. How did they approach the problem?

- Have you ever been burned by delivering code en masse? What impact did this decision have on the team?

The Done Upon Delivery Pitfall

A feature isn't complete as soon as it's been deployed; it's complete when it solves the problem it set out to solve.

Our engineering team had just had an all-star quarter. We'd delivered feature after feature at a consistent cadence—and right along with our roadmap. Stakeholders had confidence that when we committed to an initiative, we would deliver.

And yet, our company's sales slumped. After an upward trend in the past few years, we'd hit a rough patch for the third quarter in a row.

If you asked anyone from the engineering team what the problem was at the company, they'd be quick to look every direction except inward. From our perspective, we were knocking items out of the park. Customers should have been getting more value than ever out of our product, and the sales team should have had a strong pitch to bring to new leads.

However, our analytics told a different story—or rather, our lack of analytics told the entire story. The only way to even begin looking at usage metrics for the site was with our analytics tool that nobody ever used. Maybe you've seen this before—one engineer sets up a usage tracking library, and nobody ever touches it again.

We were the classic example of what is known in the software world as a "feature factory." We shipped a lot of features, but we weren't focusing on making *useful* features. We didn't even have a way to determine if features *were* useful.

The Perils of Resigning Upon Delivery

The process of creating software is in many ways similar to building a house.

With a house, blueprints are formed around the requirements gathered from the owner. In the same manner, software requirements are gathered from stakeholders and captured in the form of tickets. Both processes have a manufacturing phase. The physical demands of hammering, nailing, and pouring concrete in the construction world are akin to the mental demands of writing the code itself for digital products. The metaphor even extends to the quality assurance process. Homes have inspectors and specialists who look for deficiencies in different areas of the building, just as QA testers seek out deficiencies and explore how a new feature will affect the rest of the application.

Where the processes differ, though, is around deployment. When building a house, it becomes apparent when a feature can be classified as "done." A window's installation is done when it opens, closes, doesn't leak, and the wall around it has been smoothed over and painted. The person who installed the window and the person who uses the window can safely say the project is done. Nothing to follow up on there.

Software's lifecycle is less clear-cut. Deploying a feature by making it available to the world as part of a release is straightforward, but knowing that it is categorically "done" is a different story. This pitfall rears its head when a developer works on a new feature, deploys it, and never thinks about it again until there is something drastically wrong with it.

Suppose you're tasked with adding a retirement planner feature to a bank's website. It would be reasonable for you to add a multi-page questionnaire where a user fills out their information and submits a form. The application would then present a report that shows how much money the user will have when they plan to retire. Imagine you deploy this feature, verify that it works in production, and move on to your next initiative.

But wait—has your team honestly solved the problem you set out to solve in the first place? On paper perhaps, but there are many follow-up steps required to ensure that the feature is effective. For example, the form could have too many questions, and users could bail on the process before they complete it. It could have questions that are confusing. Maybe users even have trouble finding the form in the first place.

If any of these outcomes are the case, is your feature truly done? Although it technically would fulfill the requirements from its specification, it's certainly incomplete from a practical standpoint.

To the end users of the application, features that are unusable from a user experience perspective are just as useless as ones that have major bugs. Yet all too often, features are considered done as soon as they are available to users. Taking the "if you build it, they will come" approach to software makes a dangerous assumption that every feature is useful. Even first-rate teams still make incorrect assumptions about how users will interact with their software, and nobody gets it right on the initial attempt. Following up on usage and the user experience for new features is an essential, overlooked step in building software.

Software's Physical Therapy

While building software is like building a house, deploying software is more like performing surgery. Anyone who's had an intense operation like knee surgery can tell you that it is a long process composed of many interdependent steps. Much like the operation itself, the recovery period is crucial. An entire staff is dedicated to physical therapy—measuring how the surgery went, the current status of the patient, and how to proceed given their condition.

Just as how a surgeon should never perform knee surgery and immediately send the patient home to deal with it on their own, new software features shouldn't be released into the wild without a plan to determine how well they are working. The software equivalent of physical therapy comes in the form of analytics and iterative product planning. Once a new feature is deployed, part of the team's process should be to measure its usage and usability, as well as have a mechanism to address any shortcomings in future sprints.

In our retirement planner example, appropriately following up after deployment could mean measuring how many people are using the tool or conducting user interviews to see how easily understood the overall flow and questions are. From there, your team can iterate on the feature's design and may respond to low usage metrics by promoting the planner elsewhere on the site. If users abandon the questionnaire halfway through because it's too long, the next site update could pare down the number of questions.

It's tempting to think that these responsibilities lie beyond the purview of a software developer. It's true that depending on team structure, many of the roles outlined here are sometimes performed by product managers and designers. This could make sense in many cases, but team dynamics vary greatly from one to another. What doesn't change between teams is that a developer should strive to be *product-minded*. This means that engineers are responsible for working with designers to think about features as a whole as opposed to solely what is involved in coding them.

Even if you're not the one who's necessarily performing each follow-up task on a team, it's your responsibility to ensure that your team has the tools they need to be successful from an engineering perspective. This may range from querying a database for usage metrics to implementing an analytics library to utilizing an A/B testing framework to conduct feature usability experiments. The data that helps paint a more accurate picture of how a product is being used is a key component in crafting valuable software, and engineers play a crucial role in helping find that data.

Closing the Loop

When the company I was working at hit that rough patch, I was as much at fault as anyone for having the blissfullly ignorant attitude that product and design would take care of all usability research. I didn't make it a priority to ensure my teammates had the tools they needed to do their jobs. And frankly, it was more fun to build new features than it was to measure how effective our existing features were.

In recent years, I've found that the most success I've had is when I'm on a team that collectively takes responsibility for solving problems together. Rather than thinking of research as a "product" responsibility, everyone works together to make sure research can be done easily and in a straightforward manner.

Consider the difference in how you would approach a problem if you were asked to "build a reporting feature" versus "help users obtain useful data." It's a subtle difference in phrasing, but I would argue that the former phrasing leads a developer to rotely write a feature closely from a spec whereas the latter challenges an engineer to think critically about the problem being solved.

In the end, iterative software updates are just as important as the initial features that they accompany. A feature that provides a negative user experience but is never detected could even be worse for the application than having not written the feature at all. A developer's role isn't just to ship software, but to empower the team to prove that a change is successful.

Forging your path:

- How often does your team release big features into the wild without a plan to verify them?

- Calling to mind a feature you've worked on recently, what information was required to have confidence that a new feature was solving the problem that it set out to solve?

- What's an example of how a feature is worse than not having been written at all?

PART III: TEAM DYNAMICS

The Half Measure Pitfall

Even good ideas should not be introduced into a system without a proper plan to integrate them.

"Let's just upgrade the old reports when we come across them!" I suggested, beaming with pride at how reasonable this compromise was. In a room of contention, I enjoyed playing peacekeeper.

We were halfway through an overhaul to our reporting system. A third of our reports had been converted to use a newer, faster, shinier library. Everyone agreed that these refactored reports were simpler and just plain better overall. In the meantime, the remaining two-thirds of reports still ran using an older library. They worked well enough on their own, but we couldn't decide what to do with them.

One faction of the team thought that we should leave them as-is and not bother investing any more time into something that wasn't broken. The other faction wanted to upgrade all reports since we had already made an initial push. My proposed solution landed in the middle. I figured we could leave the old reports as-is so that we didn't need to invest any time into them immediately, tackling them piecemeal as we made further modifications in the future. This approach, I reasoned, would free up our resources while still ensuring that the conversion would happen in due time. We would put off the problem today so that we could solve it tomorrow.

Only tomorrow never came. It gradually became more and more of a pain to work with the old reports. The syntax was just different enough to be annoying to relearn whenever I updated a report every few months. I found myself explaining both flavors of reports to new team members and watched helplessly as they fumbled their way through learning the nuances of each style.

I aimed for the best of both worlds and ended up with the worst of each. I had overlooked the fatal flaw in my plan: in order for it to work, *every single report* would need to be modified at some point. This never happened in the years that followed during my time at the company. Who knows, there may still be some old reports hanging on to this day.

The Perils of Half Measures

The Half Measure Pitfall comes into play when two patterns with the same purpose are supported by a system. In particular, it ensues when a new pattern is introduced to a system as a best practice but lacks an adequate transition plan from the old pattern. What starts as an earnest attempt to improve the system instead devolves into an exercise in juggling the maintenance between two architectures.

Take the reporting example from the beginning of this chapter. Initially my plan seemed reasonable. It acknowledged the short-term need to have a better reporting tool and the long-term need to move off the old framework. However, upon closer inspection, the transition plan itself proposed a half measure in that it allowed for two patterns to coexist for an extended period of time. This dynamic can lead to major issues within the system, such as:

- **The system is less approachable.** A new member joining the team now has to learn how both types of reports work. If each of the frameworks has different dependencies on external libraries, there is a more involved setup phase to get the project up and running.
- **Time lost by context switching between frameworks.** Moving from one way of writing code to a

drastically different way requires a context switch that takes a developer longer to mentally ramp up after making the switch (see The Multitasking Pitfall for more on context switches). If a developer generally uses one framework more frequently, they will get used to the syntax and behavior of that framework. Then, anytime they attempt to do that same type of task in the other framework, they'll have to relearn that method's nuances.

- **Constant decision making for when to rebuild versus modify.** The team agreed to rewrite reports that use the old framework whenever a new feature is requested—which sounds good in theory but breaks down in practice. Although modifying existing functionality is less time-consuming and therefore easier to accommodate, it leads to a constant need for analysis with every update going forward.

- **Risk of never actually completing the transition.** By only converting a report when a new feature is requested for it, the plan to transition *all* reports won't be done until every single one has an accompanying request. What's troubling is that there's no guarantee that will ever happen, meaning the problems around approachability, context switching, and decision making introduced by the new framework will continue to live on in the system indefinitely.

Full Measures

The solution to half measures lies with more carefully weighing the decision to transition to a new pattern, and then fully committing to it once the decision has been made.

A non-coding example of a time where it may be tempting to take a half measure is the process of moving from one house to another. Say your move involves a month of overlapping leases. The transition may lead to a half measure if you decide to slowly move your belongings from the old house to the new one. What starts off as a reasonable idea leads to headaches. By taking one carload of possessions at a time, you make many more trips than if you had just rented a truck and done it in one fell swoop. Throughout the duration of the month, items are spread over each location, making it hard to locate a specific item at any given moment.

As with both coding and moving, committing to the transition phase is key. Rather than the half measure of moving some of the items over piecemeal, taking the full measure of moving all at once is far more productive.

Likewise with software where two patterns are being supported, it's better to have the rip-off-the-bandaid effect of a spike of pain over a prolonged agony. Once a team has chosen their preferred pattern, their job is then to shift away from the old methodology as fast as possible. Taking a full measure means having a definitive timeline that focuses on completeness.

Teamwork plays a key role in these transitions. What may take one person a few weeks could instead be accomplished within days if a group of engineers is focused on a singular problem. Most product managers would wince at the idea of dedicating multiple days to transition-related tasks. After all, that time could be better spent developing new features! However, as we discussed earlier, the cost to maintain two ways of doing something within a system is both high and subtle; the longer a system is in this state, the more painful it becomes.

Taking the example one step further, moving is a process where the transition cost is high in terms of both money and time. It takes time to pack everything as well as to go through the entire homebuying process of touring and closing on a house. Besides the expense of the home itself, extras like truck rental and closing costs all add up quickly. There's a reason people don't go around buying a new house every time they see a nicer one that's the same list price as their current house.

So why do developers do this with software? Simply put, it's easy to take the shortcut of ignoring the transition plan. They realize that team members are more hesitant to adopt a new pattern if they know there's a large transition plan that comes along with it. Performing the due diligence to set a team up for success in the long term is harder than getting fast results.

In the same way that moving into a new house isn't just about writing a check, developers too often think of migrating to a new framework as being something as easy as writing new code. Writing new code is the easy part, as anyone can come up with a small proof-of-concept for how to do something differently. It's a sign of developer acumen to realize that the real challenge lies in elegantly retrofitting a new pattern within the rest of the codebase.

Closing the Loop

Though my issue with the reporting system happened years ago, it's certainly not the last time I've encountered a problem made worse by half measures. These days, I agree with the faction who wanted to convert everything all at once.

If we had all focused on converting the old reports over initially, it maybe would've taken a week or so to complete the migration effort. That's a small price to pay for a burden that lasted for years instead. Ignoring the debt when planning a project just pushes it further down the road; it doesn't make it go away.

Forging your path:

- When working on a problem that involves a transition plan, think about the developer experience for someone joining the team one year down the road. How much of the "old" versus the "new" system should they have to interact with and in what capacity? What are the steps for how to get to this state in the system?

- Do you have any half measures in the main codebase you're working on? How painful would it be to take the full measure to migrate away from them?

- Consider a new experimental feature that you're not sure will be adopted by the team. As you're proposing it, picture what an ideal transition plan would look like if you adopted it and consider what off-ramp the team could employ in the event that you had to abandon the proposed feature. What does backing it out of the system look like in this case?

The Bikeshedding Pitfall

Teams can easily get caught up in discussions that lack value.

I can't tell you how long I've spent discussing project code names with teams. Or maybe a better way of phrasing this is: I can't tell you how long I've spent in *incredibly entrenched debates* with teams about project code names.

My self-indulgent comments like, "Obviously Project Phoenix[3] is the best name for this project because it represents the rebirth of our old system!" represent the type of argument I thought deserved a standing ovation. Of course, my teammates

---

[3] This example isn't specifically referring to the Phoenix web framework, of which I'm a huge fan.

had other equally cheesy project names that they favored for similar reasons.

Either way, it's been entirely too many hours. What's even more astonishing is that none of it matters.

The Perils of Bikeshedding

The term "bikeshedding" is used to describe a group of people spending too much time on decisions that provide little to no value.

The canonical example from which this term originates is that of a team tasked with building a nuclear power plant. Upon receiving this assignment, the team immediately gets caught up in a long debate about choosing a paint color for a bikeshed outside. They become so preoccupied with this choice that no other work can get done in the meantime. The end result is that the central goal of building the power plant makes no progress, thwarted by a discussion about the bikeshed's paint.

Contrived as though the origin story for this phrase may be, it does a good job of capturing the spirit of many debates within software engineering teams. These types of debates appear in every part of the software development process, from debating how to write a simple function to deciding what color a button on a page should be to narrowing down the exact text to display on a screen.

The pitfall itself occurs when teams dedicate too much time debating an insignificant topic. It's okay to establish your opinion in the first place, but going back and forth with a team member where neither person is making new points nor is anyone changing their minds doesn't help anyone.

Part of what's sneaky in these scenarios is that the team may not even recognize that they're partaking in the act of bikeshedding until they are buried deeply in a fruitless, cyclical conversation. Like an aggressive weed, bikeshedding slowly spreads until the rehashed topic somehow makes its way to being a deeply rooted point of contention.

The effects of bikeshedding may be as small as wasting a bit of the team's time or as large as a full-blown argument. Not only are time and energy spent trying to make a minuscule decision, but problems that are both minor and subjective can lead to arguments without closure, which can lead to resentment. When this happens it's particularly tragic, as by definition bikeshedding problems are inconsequential in the first place.

A deceptive aspect of bikeshedding is that the act of discourse often makes problems seem more serious than they are. The more people talk about something, the more important they think that discussion is. Upon taking a step back and reanalyzing the situation, it becomes apparent that the topic is immaterial.

Picking the Paint

Bikeshedding is a byproduct of collaboration, making it unavoidable (to a degree) because discussions are such a vital part of operating on a software development team. However, having coherent guidelines in place can mitigate bikeshedding to a mild form.

A direct way to address bikeshedding is to identify it before wasting too much time on a discussion. In the luckiest case, a team that is belaboring a discussion will have someone call out that the conversation isn't moving forward, after which point everyone will agree to quickly make the inconsequential choice in order to move on. This, however, can be easier said than done. Consider the following techniques in cases where the path forward isn't quite as evident.

Timebox the decision. Timeboxing involves setting aside a limited time to come to a decision. Suppose the team is trying to pick the name of a new project, which is of minimal importance compared to the task at hand. Rather than having an uncapped discussion on potential names, timeboxing allows the team to provide as many name suggestions as they want over the course of fifteen minutes. That could mean setting up a meeting where the end time acts as a natural cap for discussion. When a back-and-forth isn't required, it works just as well to solicit information over email or chat within a discrete time frame. This strategy is effective in cases where the team wants to make sure everyone's opinion is heard. In this example, team members have a fair and equal opportunity to contribute project names before the decision is finalized.

Identify a decision-maker. When everyone is responsible for making a decision, nobody makes it. Bikeshedding routinely occurs when a design-by-committee strategy unfolds. This could stem from fear of offending others or not feeling empowered to make a final call. Having a decision-maker sets clear expectations for the entire group. Identifying this person early allows them to proactively ask questions to inform their decision.

Conduct an experiment. Sometimes what's preventing a team from moving forward on a decision is a fear of permanence. The idea of making a poor choice that everyone else has to live with for the rest of a system's lifespan can induce fear. For example, if a team is arguing over the exact verbiage to use on a marketing page, perhaps they agree to conduct an A/B test over the course of a month to see what copy is most effective. By conducting an experiment, a problem is explicitly framed along with a solution and expected outcome. What's more is that the experiment has to take place over a set period of time, so it has a natural stopping point. Regardless of the outcome, collecting information around a problem helps inform better decisions in the future.

Get random. When something is so trivial that it barely deserves a standalone discussion in the first place, it's time to just pick something and move on. The text on a webpage that's only seen by one or two internal employees can be updated later and doesn't need an entire team of engineers going through multiple revisions as copy editors. Adopting an action bias on small things helps avoid unnecessary barriers and streamlines progress.

Closing the Loop

One hundred percent of the time I've spent debating project names would have been better spent on—well—anything else. You could make a steelman argument that an apt project name can improve its perception within a company, ultimately better setting it up for stakeholder buy-in and success. But let's be real here; as long as it's not confusing or in plain poor taste, just about any arbitrary project name is fine.

I used to get into these debates often because I'm a person who naturally cares about his work. And while passion isn't a bad thing, there's a certain "if you love something, set it free" attitude that is applicable to these types of inconsequential discussions.

If I truly wanted to help with a project, I should have helped with *the project*. Finding a good name for the project only offered the illusion of help in the form of busywork. When the project was completed, nobody said, "Boy it's such a good thing that we named it Project Phoenix, otherwise it never would have succeeded!" By reframing my mindset, I've been able to devote attention to what's most important for projects to succeed.

Forging your path:

- Think about the last time you were in a discussion that qualifies as bikeshedding. Why were the people involved so passionate about the topic?

- Are there areas of an application you're working on that are more prone to bikeshedding? Can you think of ways to avoid this pitfall in the future?

The Hill to Die on Pitfall

Trying to convince an entire team to share a personal opinion is an exercise in vain.

"I have seen the light. Django is trash; Ruby on Rails is the way." The entire room erupted with laughter, and there was an air of fanaticism inherent in the cheering that met this confession.

This quote came from a developer who was introducing himself at a Ruby on Rails meetup. Everyone knew from his tone and delivery that he was only half-serious, mostly meaning he was excited to learn about a new technology. Yet the fervor in the room was unshakeable. We had all met to bask in the glory of Ruby on Rails, and dogma dictated that only one framework could reign supreme. Anything else was lesser and, as he so succinctly put it, trash.

I can't tell you why this particular incident sticks out to me all these years later. Maybe because it demonstrates how quickly we latch on to our own belief system. There's also this idea that for something to be good, something else must be bad. Of course, this notion is ridiculous. Comparing languages and frameworks isn't a zero-sum game where one choice must fail for the other choice to succeed. Instead, they can and should live in harmony where we appreciate each for the things they do well.

The Perils of a Hill to Die On

There is one form of bikeshedding that warrants its own discussion, which will be referred to here as "the hill to die on." A hill to die on comes up in software development every time someone thinks they've found a solution they consider "the best," whether that's the best text editor, the best framework, the best language, and so on.

It's not uncommon to become entrenched in debate around these sorts of topics, especially when a team is trying to make a decision that affects everyone while navigating equally impassioned viewpoints. In these cases, there can quickly reach a point of bikeshedding (and specifically a hill to die on) if a developer evangelizes a solution so unrelentingly that it prevents others from making progress on their own work.

A question like, "What do you think about this new testing library?" may start off innocently enough, but can easily spiral into an ongoing debate that entraps and paralyzes a team. Someone who sees their testing library of choice as the only right solution may be unyielding in their opinion. It's futile to be part of a conversation where neither side has any intention of changing their mind in the first place.

Since no solution is ever going to be absolutely perfect, these sorts of discussions are extra painful. Every developer has their own experience and history that has influenced their opinions, and it's tough to appreciate an opinion about a language or framework without having worked with it directly in the past.

The Graceful Descent

There are a handful of ways software hills to die on can be resolved or avoided altogether. As always, the proper strategy depends on the context of the problem.

Seek coexistence. Many decisions don't require consensus. Topics like "What text editor should we use?" are more of a personal preference where each team member could choose individually what works for them rather than being mandated at the team level.

"Don't feed the trolls." This age-old internet saying advises ignoring internet "trolls" who are purposefully attempting to goad someone into a response. (One point of clarification with this analogy is that it is very unlikely that a team member is a true troll in the sense that their sole goal is to provoke a response. There is a difference between attempting to lure someone into an argument and simply being passionate.) The general sentiment can be applied to heated discussions as well.

When someone is firmly aligned with a particular preference, knowing when to recognize that they're not going to change their mind is a powerful trait. Arguing with them head-on is only going to make them dig their heels in further.

Dig to find the underlying point. Statements like "Ruby on Rails is the best web framework." not only lack nuance, but they're boring and useless when taken at face value. Asking someone *why* they believe one approach is superior to another sheds light on the underlying argument and provides crucial context necessary for moving forward.

If they say, "Because the best developers use it!" it's time to again dig in and inquire about what exactly they mean by "best developers." Looking beyond surface-level statements helps us find the crucial crux of intent.

Suppose their response instead is, "Because a team of experienced Rails developers can prototype a product quickly, and the majority of our team members have experience with it. As a small team, we know we can draw on examples from analogous problems on another code base we have in order to meet the tight deadline for our demo." This type of answer respects the context of the situation and considers concrete strengths of the framework. The focus has shifted away from the blanket statement of being "the best" and, in turn, highlights the relevant characteristics of the framework.

Frame the problem being solved. Every solution has tradeoffs. Framing a problem—that is, acknowledging the reality of the situation in terms of its larger context—is so effective because it helps give a sharper lens through which solutions can be viewed. What was an appropriate solution for a brand new project, for example, may not be what is feasible for the team's present situation.

For example, if a team is evaluating how to improve their testing framework, the goal of improving existing test coverage has a different set of requirements compared to the goal of accommodating more complex scenarios. A team whose goal is to expand their test coverage may choose to double down on their existing framework. Conversely, the goal of handling more elaborate testing scenarios may lead the team to the conclusion that a new framework altogether is needed. Note that neither the existing framework nor the newly proposed framework is intrinsically better; they each have their advantages within different contexts.

Going through the process of weighing pros and cons ensures that regardless of the decision, it is made with justification. Even though someone's preferred solution may not make the final cut, framing the problem takes some of the personal sting out of the process since the advocated-for benefits are still acknowledged.

Closing the Loop

Maybe I remember the Ruby on Rails meetup incident because it's the simplest example of a problem that manifests itself in my daily life as a software developer. Time and time again, even the best teams succumb to debates in which people are arguing over deeply personal preferences.

In the end, these debates not only waste time but create an inhibitive team dynamic. Rather than taking the time to appreciate diversity of thought and consider the advantages another approach may offer, too much time is spent obsessing over a self-appointed winner in a made-up game with no stakes.

What do all of the suggestions for combatting a hill to die on have in common? Empathy. The sooner you're able to see an argument from another perspective, the sooner you can help your team move forward. Nobody is right 100 percent of the time; by learning to have an open mind, you open the door to creating better solutions than any one developer on a team could come up with individually.

Forging your path:

- What can be learned from someone who feels extremely passionate about a tool or technology?

- Do you ever find yourself evangelizing a certain technology? Part of what makes a hill to die on so engaging is feeling like you need to convince someone of your opinion. Next time you're in one of these discussions, assume that you already know that the person you're talking with is definitely not going to use it. How does this change how you would explain it?

The Overaccommodation Pitfall

Left unchecked, exceptional practices that are meant for dire circumstances become commonplace.

It took me a long time to notice that my team's workdays kept getting longer. Our hours were steadily increasing here and there to work on "just one more" feature. What started as *sometimes* working into the evening led to *always* working into the evening. Weekends were once sacred, but the same pattern followed there, too. Working one Saturday paved the way for future Saturdays. By the time I knew it, it actually felt strange to have a full weekend *without* work.

Yes, I had been in a similar boat in the past in my recounting of The Hardcore Mentality Pitfall. But this was different.

When I had a hardcore mentality, the pain was self-inflicted. It was an attempt to make up for my own shortcomings by buckling down on a problem and never looking up.

This time, the rigorous schedule was an affliction that impacted the entire team. The herd mentality we had was that we were always in "on" mode. Every problem and feature we encountered was given drastic priority to the point where it felt like we had our foot on the gas with no intention of ever braking.

They say that if you throw a frog into boiling water, it will jump out right away. But if you place a frog into a pot with room-temperature water and slowly heat up the pot, the frog will boil to death. The poor fellow never notices the gradual buildup toward his impending doom.

At work, we would quickly succumb to the levers that turned up the temperature. New features and new deadlines turned the dial. That prototype for the CEO's demo or that new feature that needed to be built by the end of the quarter felt like they'd mark the point where we could jump out of the water. Instead, we were greeted with a fresh pot for the next presentation and the next quarter.

At that time in my career, there seemed to be no shortage of pots or water. As for our performance, we were stuck in a vicious cycle: the harder we worked, the worse our results.

The Perils of Overaccommodation

There is a fine line between accommodation and overaccommodation. Accommodation is doing work and doing it well. Overaccommodation is deviating from the norm by dedicating extra attention to a specific task. It comes at the expense of resources like time or priority.

Overaccommodation is a deceptive pitfall because going "above and beyond" is praised within a team. Those who bend the rules to make things happen do so with good intentions, but in their pursuit of productivity, healthy processes are lost on a team. Most commonly, overaccommodation comes in two flavors:

1. Working Long Hours

The teammate who consistently works overtime to accommodate feature crunch is viewed as a hero. If management cites sixty-hour work weeks as a factor for a developer on a team earning a promotion, that signals to other developers that a sixty-hour week is what's expected by management.

Praising overaccommodation only stands to establish a new norm. Before you know it, the entire team of developers is working overtime. Or worse yet, some developers hold out and maintain a normal workload, only to be resented by the overaccommodating developers who are working extra hours.

2. Urgent Requests

The engineer who regularly skirts around a team's established process to quickly take care of seemingly "urgent" requests also suffers from overaccommodation. Stakeholders who are used to having one-off requests handled immediately quickly lose sight of what's genuinely imperative. They begin to file every request as urgent because it's more convenient to have it handled right away. Taking on just any new task immediately means that it's being misprioritized. Expectations become warped, leading to practices that are unsustainable.

Acknowledging Exceptionalism

Rather than solely praising the exceptionalism that comes along with overaccommodation, it's important for a team to acknowledge and discuss it in a team setting. A team should take a moment to identify what happened, what the expectation is for instances like this moving forward, and whether any additional action is required.

Let's consider an example of a teammate working long hours. Suppose a team member works sixty hours in a week because the site's payment system is down and direly needs to be fixed. During the team's retro, they acknowledge that this is authentically an exceptional instance where core functionality was disrupted. Then they reiterate that forty hours a week is generally what's expected from a single team member.

In terms of follow-up actions, one possibility is that the team member who worked extra now embraces a more flexible schedule for a few days. In terms of the overall application, maybe the team decides that more tests and notifications should be put into place to avoid similar problems in the future.

A comparable example with a different outcome would be a team member who regularly works sixty-hour weeks. During the team's retro, the team asks if this is due to overly aggressive deadlines or expectations. Surprisingly the developer responds that they are really just passionate about their work and enjoy putting in extra hours. Although no action may be required, reiterating what the expected amount of work is for members of the team helps maintain sustainable practices going forward.

When it comes to working on urgent requests, a developer may have a desire to get last-minute tasks off their plate. But finishing them quickly doesn't remedy the underlying issues. The question that needs to be asked is *why* a stakeholder can put something directly onto a developer's plate in the first place. Instead, new tasks should go through the proper channels, such as making a request through the product manager. There are larger systemic issues at play in cases where these channels are skipped.

Performing new tasks immediately just because they were most recently requested isn't worth the additional cost. It's incredibly rare for one-off requests to actually be pressing. Attempting to immediately address *every* task that comes in gives unfair preference to recent requests and neglects pragmatically prioritizing them.

Closing the Loop

Why do teams always try to bestow such high worth upon deadlines? Well, because sometimes things *are* that important. Maybe the startup you're at really *does* depend on this next pitch to raise a level of funding. It's right there in the name, after all—the "dead" in deadline.

The problem is that not every problem belongs in the world-ending, all-or-nothing category. If you find yourself hopping from one dire initiative to another, something is wrong. The company is either hanging on by a thread or (much more likely) is simply mis-prioritizing work. Neither case is ideal, nor will they result in sustainable feature development over the long term.

Making sure that exceptional practices stay exceptional takes a team with diligence and healthy communication. A team with these qualities then has the ability to set expectations and alignment both internally as well as externally with other teams.

Forging your path:

- Are there any practices on your team that have formed as a result of overaccommodation?

- What types of tasks are considered urgent by your team? What constitutes your team's definition of urgency?

- When was the last time someone on your team worked a fifty-hour week? Could this situation have been avoided, and if so, how?

The Default to Yes Pitfall

Great engineers don't just blindly implement every feature they're given. Instead, they collaborate with teammates to craft the solution that is the right fit for the problem at hand.

"So would you say this is a crucial part of the user experience?" I inquired, spoon-feeding the justification to our designer.

"Yes, absolutely. I understand it may be harder, but it would be incredibly important to our users," he replied.

That's it. That's all the validation I did for justifying a massive new requirement for our project, despite knowing that I had just sealed my fate and had agreed to take on much more work.

Note what I *didn't* say in my response; to my own peril, I didn't explain that building one feature meant forgoing another. I didn't request an investigatory task to examine how feasible the proposal was. I didn't ask for clarification as to what problem we're solving by building this solution.

The feature being discussed was enabling auto-save on a set of forms that users filled out as they created new projects in our system. There were maybe five screens or so, and the problem we were looking to avoid was having the user lose data if they abandoned their form without clicking on the "Save" button on the screen.

I begrudgingly implemented the feature as requested, only to see the code unfolding before me gradually unravel into a pile of garbage. Sometimes being a full-stack developer means you get to mess up both the front and back end within the same feature. This was one of those cases; I managed to make the code everywhere more convoluted in a race to prove that this request was possible.

Ironically, I had bent over backward to accommodate a feature that wasn't particularly important in the first place. Our application had maybe a handful of projects being created daily, if that. How likely was it that during that time, a user would abandon their form mid-wizard? And in the rare case that did happen, how much data would they actually lose?

Much like the solutions discussed in The More Code is More Value Pitfall, I would have benefitted from taking a step back to think about the context around our problem. Then I would have been reminded that our app was a niche product that was only used by a handful of power users at each client's company, not a mass-market product where users were creating thousands of projects each day. But one key difference in this case was that I *didn't* think I was adding more value by writing more code here. In fact, I could have easily saved myself a lot of time and effort by knowing how to push back on the feature's requirements. By having better communication, I would've made both my own life easier and written simpler, more elegant code for future contributors.

The Perils of Defaulting to Yes

Perfectionism is a mixed blessing. It can lead to creations of beauty worthy of immense pride. Despite its charisma, perfectionism also requires valuable resources like time and effort.

The implications that perfectionism has on software are not surprising. Attempting to make software impeccable is an impossible task, as there is eternally more to be done—more edge cases to address, more performance to optimize, and more polish to add to a design.

Perfectionism emerges as a pitfall by "defaulting to yes" when working on a task. Defaulting to yes means that a developer approaches every single feature request with the intention of fulfilling it verbatim. In doing so, they treat every design detail and technical requirement as gospel.

Defaulting to yes is alluring to a software developer because of how straightforward it is. By working on a feature exactly as requested, there's not a lot of keen analysis required. But product managers and designers don't have complete information on how software is to be written. Within a ticket description, what's being requested is one way to solve a problem; but it doesn't mean that it's the only method to solve that problem. It's up to the engineer to help distill what's being asked for and offer a solution that elegantly balances trade-offs within a system.

Furthermore, by spending more time than is needed on a task, the team inherently must compromise on other tasks. By bending over backward to accommodate intricate technical requirements or designs, the engineer makes a system harder to maintain and more brittle going forward. By engaging in arduous tasks, the developer drains their own energy reserves on a problem that isn't worth the hassle.

An Eye Toward Refinement

We should always strive to approach our work with a critical eye toward refinement. A task presented without engineering input can be thought of as a draft which can and should evolve as new information is uncovered.

It's enticing to think the act of deviating from the original specification is a sign of weakness—that whoever first wrote the spec lacked due diligence and now has to flip-flop. Upon closer inspection, the opposite is true. Staying beholden to the first draft of a ticket reinforces rigid requirements for the sake of pedantry, whereas refining the request from a ticket cultivates the necessary give-and-take that comes along with software development. It promotes fluidity and agility, leveraging the information the team has at the time.

For example, imagine this scenario. A team is working on a web application that their clients use internally within their organization and externally to share reports with other companies. A feature tasked to this team is to "white-label" the application. White-labeling allows clients to customize the theme of the entire site to match their preferences. As a starting point in this scenario, a client can pick the precise colors of the background of the page, the font color, and the font family of each page's content.

An engineer who defaults to yes jumps right to a solution, "We can create a database table with an entry for each company. Our application layouts and stylesheets have a single entry point for these types of styles, so we only need to change a few lines of code. All we need to add is a settings page for the company. This should only take a week to implement!"

But there is a monumental difference between whether the team *can* do something and whether they *should* do something.

Take a moment to reflect upon the potential enormity of the

impact of this request. Allowing customers to pick the colors for *any* page on the site has testing implications for *every* page of the application going forward. What happens if an image embedded on the website has the same color as the background a client picks? What are valid choices for button backgrounds if the text could be any color? Entropy not only increases for the engineering and QA teams but for the team that writes content as well. Allowing for different font families means there will assuredly be issues where a message that fits on one line with one font overflows to two lines with another font. The list goes on and on, and the product has been forever altered by this seemingly innocuous request.

Having an eye toward refinement helps explore ideas quickly and then iron out their details. An incredibly useful concept that can foster refinement is the Pareto Principle, which is the idea that 80 percent of results are yielded from 20 percent of effort (for this reason it's also known as the 80/20 rule). The inverse implication is that 80 percent of effort is spent on achieving the remaining 20 percent of results.

Under the Pareto Principle, there are often alternate software approaches that can provide the majority of the value while minimizing effort. The key is finding the solutions where compromises are least impactful. It's feasible that a solution does introduce some new edge cases, but they're edge cases that are so unlikely that they can be ignored.

Returning to the white-label example, there are multiple ways an engineer could suggest downscoping the original specification while maintaining the spirit of the request:

- **Limiting the feature's breadth.** If the underlying goal is to make it so clients can share reports with other companies where the page uses the client's branding, then only these external reporting pages need to support white-labeling. Having a subset of pages that support custom colors and fonts is tremendously more maintainable than every page on the app. In applying the feature to minimal pages based on the use case, concerns around development, testing, design, and content authoring are greatly mitigated.

- **Reducing complexity.** In this case, reducing complexity could mean hardcoding a subset of color palettes for a small set of customers. There is a possibility that this feature is only used by a handful of high-priority clients, and in the meantime, handling those customers in a one-off capacity is more elegant than jumping right to a solution that works for any customer.

- **Revisiting which parts of this request are essential.** Do customers *need* to be able to pick their font family, or was that a brainstorming idea that made its way into the requirements? Even if customers say, "Yes, this would be nice," that's not the same as them being willing to pay for an additional feature. Nor does it prove that the feature is a fundamental part of their experience. Cutting out unnecessary facets in the first pass of a feature increases the odds of success as both the code and solution itself become more concise.

An advantage of downscoping is that smaller changes are easier to revert later on. Consider what happens if white-labeling is found to be a feature that customers don't actually use. It's an order of magnitude easier to remove the downscoped version that only lived on a handful of pages than it is to revert the full-blown version that was woven into the entire site. *Widespread changes warrant confidence in their justification.* Until the importance of a feature is proven, it's better to confine the repercussions of such far-reaching changes.

Most problems have an 80/20 solution that accomplishes the same core intention of a task but that is simpler and involves less engineering effort. Paring down requirements means distilling a request to a simpler form, naturally prioritizing what is so useful about a feature. But this only happens when you're able to avoid defaulting to yes. Being able to recognize and suggest such solutions is part of what separates great software developers from mediocre ones.

Closing the Loop

If I'm being completely honest, I'll admit that there was a level of pride that got in my way as well when I defaulted to yes in the past. I always wanted to *prove* that I could handle the more difficult version of a given request. I figured that the best developers were the ones who could handle the hardest features. It didn't occur to me that developers could actually help stakeholders by pushing back on requests in order to balance code simplicity with the value provided to a customer.

Another idea that pushed me toward my original mentality is "What's best for the user is best for the product." Although it's true that products are built in service to a user, that's not the same as building every single feature that may provide *some* level of utility to a user.

Being a great software developer means being able to say no to features that may be useful to a user but impractical within the given constraints of a system because, as we know, every feature comes with an opportunity cost. Ultimately, if a change is important enough, it will make its way back into the backlog. It's rare for software to really only have one chance to solve a problem. Teams who aren't working on mission-critical systems have the luxury of being able to evolve and adjust accordingly.

Forging your path:

- In your team's current process, when would be the best time within the product development lifecycle to have engineers help weigh in to find 80/20 solutions?

  - Is it during the initial ideation phase for solutions, or after a design has been created? There are no right or wrong answers, as this can vary from team to team and problem to problem.

- When working on a big new feature, how confident is your team that the proposed solution is the correct solution?

  - How much research has been done beforehand to verify this?

  - What are other opportunities for research that the team could implement in order to increase confidence for other features going forward?

The Being Right Pitfall

Having amazing feedback is ultimately
moot unless it's delivered in a way that the
recipient will be receptive to.

I've been wrong a lot in my career. From specific coding to
higher-level architectures to designs, I'm not ashamed to admit
that I've held ill-informed positions on just about any topic
you'd encounter as a software developer.

When people have pointed out that I've been wrong, I've
had all kinds of reactions. When someone challenges my view
on something, sometimes it's very easy for me to change my
mind on the topic. Other times, a purely emotional response
takes over where it's almost impossible for me to change my
mind even if deep down I know that the other person is right.

What separates advice that is easy to take versus advice that's hard to take? For me, it all comes down to the presentation of the advice itself. In fact, I would contend that the tone of delivery is the single biggest factor for determining if the advice you give someone will be accepted.

When I had this realization, I took some time to reflect upon my own advice.

At my best, my feedback was empathetic to my teammate's situation, and I made it a point to balance being direct and understanding in how I presented it. Rather than aspiring to simply be right, I often viewed my role as someone who could help shed light on more information in a way that let others arrive at their own conclusions. This often involved asking probing questions to empower code authors to weigh tradeoffs on their own.

At my worst, I could be too blunt or dismissive of someone's approach. In code reviews, I would matter-of-factly outline what was wrong without any additional tact. I was so focused on providing the feedback itself that I didn't consider how it could've come off as cold. Additionally, I wasn't fostering a team environment where people would feel comfortable getting help from others.

The Perils of Being Right at All Costs

Giving feedback is an innate part of software development.

Particularly in code reviews, there is ample time to advise others. Curiously, this is a skill that we seldom focus on as developers—with intention, that is. Yet, when done well, providing useful assessments is a potent trait that helps individuals amplify their impact on a team.

When we think of giving advice, we tend to think of being *right*. For example, suppose a junior developer on a team has submitted a code change that is being reviewed by a senior member with more experience and insight. That senior member may think that their role is to be right about as many things as possible by finding as many mistakes as possible.

This is a dangerous notion, as it ignores the underlying intent of code review. The review exists to improve both the quality of the code and the author's skills. The senior developer may be quite capable of leaving 100 comments on the review that dissect every nitty-gritty aspect of the code. Even when the comments may all be valid concerns, have they really created a net-positive experience for the junior developer? Will this interaction make the junior developer more likely or less likely to go to the senior developer for guidance in the future?

Being right isn't enough when it comes to advice. Poorly crafted feedback becomes a pitfall when someone mistakes being right as their only responsibility as a team member and staunchly puts it above the general well-being and psychological safety of others on the team. Someone who solely focuses on being right tends to make these types of mistakes with communication tools:

Too brief of feedback. Succinct comments like "This should be done X way instead," on their own are not very insightful. Tone aside, it doesn't explain *why* the change needs to be made. Even if the author makes this change, they only know not to repeat that exact same mistake going forward and won't have a basis for identifying similar situations in the future or building upon the knowledge that the reviewer intends to convey.

In many ways, this ends up being a "give a man a fish" scenario. The reviewer gives the fish in the form of a quick fix that's only needed at the moment, robbing the author of the opportunity to learn to fish through learning the underlying rationale that would help set them up to make improvements on their own in the future.

Too much feedback. Often combined with brief feedback can be a barrage of feedback. Giving someone 100 comments about every single thing they could have done better is as overwhelming as it is shortsighted. Beyond being discouraging, it opens the door for them to make other mistakes in trying to address so much feedback at once.

Part of the reason for code review in the first place is to make *actionable* proposals. Drowning someone in dozens of points leads to feedback composed of mixed levels of impor-tance. This dilutes what points are crucial and makes it harder for the recipient to take action.

Tone deafness. It's notoriously hard to decipher tone from text in isolation, so a blunt "This should be done X way instead" isn't automatically bad depending on the context. Though something along the lines of "Why are you writing this code this way? It should be done like X!" comes across more harshly as they might as well have written "You fool!" at the end.

Before giving feedback, ask yourself one thing to help you avoid tone deafness, *Is the way that I've phrased this going to be something that my teammate is receptive to?* Even the best advice in the world will fall on deaf ears if it is insensitively constructed. It's a shame when this happens, as both sides lose in a case where a true win-win is within reach.

Digging in your heels. Those who are resolute in being right find themselves digging their heels in with every opinion they give because they're convinced that others would be better off if only they adopted *their way* of doing things. However, if every preferred solution to a problem must be enforced at all costs and without the solicitation of additional context or opinions, others on the team will soon become either numb to or exhausted by the feedback. When everything is a top prior-ity, nothing is.

The reality is that deadlines, fatigue from making too many changes at once, and differing opinions all provide roadblocks to implementing every bit of criticism. Constructive discourse relies on all sides having a natural back-and-forth. Code thrives when multiple parties are able to weigh in on a solution, and this can't be done if someone is digging in their heels.

From Right to Helpful

In order to truly escape the depths of The Being Right Pitfall, try going back to the idea of a beginner's mind while shifting your perspective to one of service. Instead of asking, "How can I prove that I'm right about this?" you should be asking, "How can I be helpful here?"

Letting go. Though it can be easier said than done, it's okay to let go of being right. Even if you're absolutely convinced that your way is the right way, allowing room for others to experiment with ideas and concepts on their own is a great gift to bestow upon another developer.

As developers, we don't get better by having overbearing supervision at all times. We need slack and room to play around with ideas on our own sometimes. The old saying "Experience is the best teacher" applies to code as much as it does in many other areas of life. There's something to be said for the freedom to learn things on our own, especially if that includes the freedom to fail in a safe and contained environment.

In a low-stakes environment or a trivial change, try not to feel hesitation in letting go of your closely held opinion. The time to be staunch is better saved for critical issues with dire consequences. Naming a CSS variable versus a bug that ripples throughout an order system warrants very different levels of severity when it comes to putting your foot down. It takes maturity and foresight to state an opinion but let another proceed with what you consider to be the wrong path.

Notably, letting go of being right is different than "being wrong." Letting go of being right doesn't mean suddenly embracing giving poor instructions. Rather than thinking of every scenario with a right-versus-wrong mentality, it's better to acknowledge that someone may need to approach a task differently to take the next step in their journey.

Communicating what needs to be said. A useful lens through which giving advice can be viewed is to consider "what needs to be said" as an alternative to being right.

Focusing on being right above all else is one-dimensional. It cuts straight to the chase without showing how you thoughtfully arrived at that conclusion. It's a rigid take that leads to a preference for dogmatic principles.

Someone who is good at being right is also typically good at making their point known. "This function isn't performant" is an example of feedback that someone could leave on a code review. It is a straightforward and likely valid statement, but if the person on the receiving end is looking for palpable take-aways, then this statement isn't helpful.

Saying what needs to be said includes tailoring a message to its recipient. This doesn't mean that each message must be carefully crafted and mulled over for minutes on end before hitting the send button. It *does* mean that it's important to make a habit of considering who the audience is when it comes to giving feedback.

Let's look at how to improve the feedback "This function isn't performant." Here, saying what needs to be said would involve identifying the problem with enough context to fix it now and in the future. To boot, it would do so in a tone that makes the author motivated to implement the feedback.

One example of improvement for this feedback would be to say, "This function will take a long time to execute with a large collection of thousands of elements since it's making multiple database queries for each element. Because the majority of users will meet this condition, I've marked this as a blocker for this code change."

Yes, this new suggestion is more of a mouthful, but it meets the criteria of saying what needs to be said. It clearly identifies the problem by explaining the issue broadly enough that it could apply to other situations in the future. Along with that, it justifies why the change is needed and leaves no doubts as to what the expected outcome of this feedback is. Tone-wise, it is direct while avoiding any accusatory language.

The difference is substantial.

Closing the Loop

I've become more intentional with the feedback I give to others. The biggest difference is that now I make every effort I can to be empathetic.

I actively make an effort to understand opposing viewpoints when they arise. Sometimes I'm convinced of them. Sometimes I'm not convinced of them but am ok with that person making their own decision. Sometimes I give the advice that I think that person needs to hear.

Giving great feedback is an art of its own. It requires a delicate balance of messaging, tone, empathy, and overall tact when it comes to delivering a message. Once mastered, it lays the foundation for the kind of communication required to have a positive impact on an entire team, making a developer more effective than they could ever hope to be on their own.

Forging your path:

- Outside of code reviews, what venues are there for giving feedback on your team?

- Think back to some excellent input that you've received and taken to heart. How was it framed and communicated?

- What's a topic that you feel strongly about, but can see the other side's opinion on?

Conclusion: Walking the Path

Navigating pitfalls in this work is a difficult craft. It would be convenient to have strict software development rules that we could follow in order to succeed, but that would also be quite boring, wouldn't it? As developers, we shine when it comes to problem-solving and critical thinking.

I mentioned in The Fear of Failure Pitfall that experience is the best teacher. As I reflect upon my own journey in software, I've found this notion to be absolutely foundational to embrace early on, as it opens the doors to so many other lessons. Nothing sears itself into your memory and influences your future choices quite like making a bad decision and directly seeing its consequences.

That's why this book is filled with warning signs, cautionary tales, and metaphorical safety belts when it comes to thinking about software. If you don't find these pitfalls, they will come and find you in one way or another. But fear not, you already have the advantage. Awareness is half the battle, and now you'll not only be able to see these pitfalls coming, but you'll also be ready to tackle them head-on.

Each time you conquer these bouts, I hope you take a minute to do two things. First, pause and celebrate your victory before you move on to the next challenge. It takes grit and intention to do things differently and see around corners, so give yourself credit for developing and nurturing that strength of mind. Second, I encourage you to share what you learned with someone else. When I began my career all those years ago, I never thought I'd be writing a book about how to navigate pitfalls in this field—but here I am. All my mistakes have taught me there's a better way—a way that impacts my experience as a person doing this work, the team dynamic at play, and the quality of the output. I care about all three of those things, and if you're reading this, odds are that you do, too.

Writing software is a complex, (sometimes) infuriating, rewarding, lovely profession. I truly hope that the lessons here provide a guiding light as you chart your own journey. Just know that even when the going gets tough, there's only one thing to remember... keep calm and code on.

Your Voice Matters!

In the spirit of sharing, I have an ask: What did you think of this book? Were there pitfalls that struck a little too close to home for you? Are there pitfalls that you think I've missed? Whatever the case, please leave me a note at hi@pitfallsbook.com.

And of course, if you've enjoyed this book, could you share it? Post it to social media, start an engineering book club at work (hey, you've already read it so you might as well get some credit!), or gift it to your friend who is thinking about getting into software development. Sharing what we learn keeps us connected to one another, to the work itself, and to the future of this industry we love so much.

Acknowledgments

This book would not have been possible without the people listed here. I'm beyond grateful for each and every one of them.

For my wife, who has done more than I can write here. Be it on nights, weekends, or any other time I could find a moment's solace to write, she stepped in and made it possible for our lives to continue during this endeavor. Did I mention that she gave *birth* to our first child while I was halfway through writing this book? Even then, she was committed to helping me see this through during the chaotic year to follow in our lives. I am fortunate to have her by my side.

For my daughter, who is still an infant at the time of writing this. Her smile alone makes me want to become the best version of myself. Her multi-hour naps are a blessing I will never take for granted (not to mention a blessing that allowed me to get through the last couple of editing revisions).

For Keith, as a colleague, mentor, and all-around generous human being. His dedication to excellence in software continues to inspire me years after working together. Anytime I talked about someone who has a positive force on an entire team in this book, I thought of him.

For Dan, who's humored my many side projects over the years and is always keen to lend thoughtful insights. He was one of the first people to read this book in its most raw form and provided the encouragement and feedback that helped shape it in many crucial ways.

For Adam, who literally makes me rethink my definition of a friend in that he goes above and beyond with any favor I've ever asked of him. Within this book, I can only describe him as having the "Midas touch" of advice. His ability to take a rough idea and bring out the best in it is unrivaled.

For Josiah, whose support both professionally and personally means the world. His compassion during some of the most difficult times of my life is an act of kindness I will forever be thankful for. I'll be lucky if I ever have another manager like him in my career.

For Sarafina, who helped turn this book from a side project into a professional endeavor. I am in awe of her natural selflessness and eagerness to help me overcome many of the biggest roadblocks I encountered along the way.

For my editor, Jessica, whose guidance has far exceeded what I could have hoped for. She has a knack for drawing out the words I *want* to say rather than what exists on the paper in front of her. Her patience and empathy have made her a pleasure to work with during this entire process.

For my parents and sister, who sacrificed so much to put family first and who have always believed in me. I will never forget when my father read my first published article and, showing his unwavering support, said, "I have no idea what any of that meant, but I read the whole thing!"

About the Author

In over fifteen years of writing software for a variety of organizations, Alex has learned a lot about not only this industry but what it takes to succeed within it. Whether it be at a large enterprise or one of the startups he's worked at (two of which were part of successful acquisitions), he has first-hand experience with how to build applications and just as importantly, how not to.

One thing that hasn't changed over the years is that Alex is quick to empathize with those starting their journeys in software. In fact, this affinity for mentorship and knowledge sharing is a big motivation for this book. Alex is quick to point out that his success as a software developer is rooted not in a natural coding ability but rather in his thoughtful and strategic approach—one he's excited to share with his readers.

In the big picture, Alex is a passionate proponent of work/life balance as a tactical advantage, a believer that soft skills in software are just as important as hard skills, and a champion of the idea that software can be mastered by anyone with the right mindset.

Outside of work, Alex enjoys finding the right meme for every situation, having a dad joke at the ready at all times, and telling anyone who will listen about the time he was awarded Evaluator of the Year in his local Toastmasters group. When he's not thinking about code, chances are he's planning his next run in Slay the Spire.

Alex continues to work in the software industry and lives in Texas with his family.

www.ingramcontent.com/pod-product-compliance
Lightning Source LLC
Chambersburg PA
CBHW030507210326

41597CB00013B/828

9798990667211